Regional Policy and Planning in Europe

Paul Balchin and Luděk Sýkora, with Gregory Bull

London and New York

First published 1999
by Routledge
11 New Fetter Lane, London EC4P 4EE

Simultaneously published in the USA and Canada
by Routledge
29 West 35th Street, New York, NY 10001

© 1999 Paul Balchin and Luděk Sýkora, with Gregory Bull

The right of Paul Balchin and Luděk Sýkora, with
Gregory Bull to be identified as the Authors of this
work has been asserted by them in accordance with
the Copyright, Designs and Patents Act 1988

Typeset in Garamond by
J&L Composition Ltd, Filey, North Yorkshire
Printed and bound in Great Britain by
Redwood Books, Trowbridge, Wiltshire

British Library Cataloguing in Publication Data
A catalogue record for this book is available
from the British Library

Library of Congress Cataloging in Publication Data
Balchin, Paul N.
  Regional policy and planning in Europe/Paul Balchin and
  Ludek Sykora, with Gregory Bull.
    1. Regional planning–Europe. 2. Europe–Economic policy.
    I. Sykora, Ludek, 1965–   . II. Bull, Gregory H. III. Title.
  HT395.E8B35   1999
  338.94–dc21   98–42340

ISBN 0–415–16009–X (hbk)
ISBN 0–415–16010–3 (pbk)

Contents

Figures

Tables

Abbreviations

A	Austria
B	Belgium
BG	Bulgaria
BMBau	Bundesbauministerium (Germany)
CAP	Common Agricultural Policy
CEC	Commission of the European Communities
CEE	Central and Eastern Europe
CH	Switzerland
CI	Community Initiatives
CoR	Committee of Regions
CZ	Czech Republic
D	Deutschland (Germany)
DDR	German Democratic Republic
DK	Denmark
DoE	Department of the Environment
DETR	Department of Environment, Transport and the Regions
E	España (Spain)
EAGGF	European Agricultural Guidance and Guarantee Fund
EBRD	European Bank for Reconstruction and Development
EC	European Communities (or Community)
ECB	European Central Bank
ECE	East Central European
ECSC	European Coal and Steel Community
ECU	European Currency Unit (also ECU, ecu)
EEA	European Economic Area
EEC	European Economic Community
EFTA	European Free Trade Association
EIB	European Investment Bank
EMU	Economic and Monetary Union
EMI	European Monetary Institute
ERDF	European Regional Development Fund
ERM	Exchange Rate Mechanism
ESF	European Social Fund

ESDP	European Spatial Development Perspective
EU	European Union
EUR 12, 15	European Community according to the joining sequence of Member States
Euratom	European Atomic Energy Community
EUROSTAT	Statistical Office of the European Communities (also Eurostat)
F	France
FIFG	Financial Instrument for Fishery Guidance
FURs	Functional Urban Regions
GDP	Gross Domestic Product
GDR	German Democratic Republic
GR	Greece
I	Italy
IRL	Ireland
ITT	Information Technology and Communication
L	Luxembourg
NL	The Netherlands
NUTS	Nomenclature des Unités Territoriales
OECD	Organisation for Economic Co-operation and Development
PHARE	Pologne et Hongoire assistance pour la Réconstruction Économique
PL	Poland
R	Romania
S	Sweden
SB	Sorbia
SEA	Single European Act
SEM	Single European Market
SF	Finland
SK	Slovakia
SME	Small and medium-sized enterprise
SWE	Sweden
TACIS	Technical Assistance for the Commonwealth of Independent States
TEN	Trans-European Network
TGV	Train á grande vitesse
UK	United Kingdom of Great Britain and Northern Ireland
UKR	Ukraine
UNECE	United Nations Economic Commission for Europe
VAT	Value Added Tax

1 Introduction

This book is intended to facilitate the study of regional policy and planning, both within the context of the European Union (EU) as a whole and within the individual countries of Europe. During the last decade of the twentieth century, widespread concern was expressed about the integrative role of regional policy and planning in furthering the economic, social and political coherence of Europe. Within the context of the impending formation of the Economic and Monetary Union (EMU) and the enlargement of the EU, there was much debate over whether national and regional disparities in living standards and unemployment would be widened or narrowed in the early twenty-first century. Arguably of equal interest were constitutional developments, which were paving the way in many of the countries of Europe to various forms of regional government and regional planning. Whereas it had long been recognised that the competence for regional planning on a continental scale should appropriately be assumed by a supra-national organisation rather than by a loose collection of national states, it was becoming clear that regional planning within individual countries could be handled more effectively by the regions themselves rather than by the centralised state. However, the formation of regional tiers of government and the development of various forms of regional planning were proceeding at a different pace from country to country. Only where cross-border planning was undertaken was there a possibility that a cohesive approach to regional planning would emerge – short of a uniform system of planning being imposed across the EU. Other important considerations included the degree to which investment in transport, information technology and energy would help or hinder the improvement in living standards in the peripheral and other disadvantaged regions, and the extent to which environmental improvement had an impact on regional development. Last, but not least, the economy and environment of urban areas are of considerable concern to policy-makers, particularly since 80 per cent of the population of the EU live in towns and cities rather than in the countryside. Urban areas, however, are inextricably incorporated into the economy of their regions, and consequently their problems often require regional rather than discretely urban solutions.

The Economic and Monetary Union and the enlargement of the European Union

On 25 March 1998, the European Commission confirmed that eleven Member States of the EU had, in effect, met the economic convergence criteria of the 1991 Maastricht Treaty and were thus eligible to adopt a single currency (the euro) in 1999 and become founding members of EMU. Of the other EU states, Greece had sought membership of EMU but had failed to comply adequately with the Maastricht criteria, while, for their own reasons, the UK, Denmark and Sweden were not among the first wave of applicants. However, although the membership of Germany, France, Italy, the Netherlands, Belgium, Luxembourg, Ireland, Spain, Portugal, Austria and Finland was subsequently endorsed, commentators were quick to point out that in six of these countries (and most notably in Italy and Belgium), debt ratios were in excess of the Maastricht Treaty requirement of 60 per cent of the gross domestic product (GDP). All eleven members, nevertheless, conformed with the requirement that budget deficits should not exceed 3 per cent of GDP, and were broadly on target in respect of long-term rates of interest and levels of inflation. Although, in most countries, progress towards meeting the Maastricht criteria was evident over the period 1993–98, the European Monetary Institute (EMI) (the predecessor of the European Central Bank) was concerned about whether this would continue once EMU was under way in 1999 and in the early years of the new century.

In the larger Member States, there were major economic problems that needed resolving before the success of EMU in its early years could be assured. With the unification of Germany, the ratio of its national debt to GDP soared from 41.5 per cent in 1991 to 61.3 per cent in 1997, but although its budget deficit was forecast to be as low as 2.5 per cent in 1998 this might still have been too high to have curbed the debt ratio significantly. Substantial fiscal measures seemed necessary if the debt ratio was to be reduced to 60 per cent or below in a reasonable period of time. Although France (together with Luxembourg and Finland) complied with all the eligibility criteria for a single currency, its debt ratio had risen sharply from 35.5 to 58 per cent of GDP, 1990–97, mainly because of increased public spending on unemployment, health and pensions. To prevent the debt from rising further, its budget deficit needed to be decreased from 3 per cent of GDP in 1997 to 2.5 per cent according to the EMI. Italy was notable in having a debt ratio of 118 per cent of its GDP in 1997 – almost twice the Maastricht level of 60 per cent, but, because of far-reaching budgetary measures, the public deficit had been reduced dramatically from 9.5 per cent of GDP in 1993 to 2.7 per cent in 1997. However, according to the EMI, large and persistent budget surpluses will be necessary to reduce the debt ratio within an appropriate time. If, for example, Italy ran a recurring budget surplus of 3 per cent per annum, then a government debt of 60 per cent of GDP could be achieved by the year 2007. Although not among the first-wave applicants of EMU, the United Kingdom, nevertheless, would have been eligible for entry had it applied, since its debt ratio of 52 per cent of GDP in 1997 was well within the Maastricht requirement, and so too was its public deficit of only 1.9 per cent. Membership, however, would have necessitated a marked lowering of the UK's long-term interest rates – a downward movement incompatible with its aim of ensuring that the rate of inflation remained within target. However, on the assumption that the UK is able to continue to adhere to the Maastricht criteria and maintain stable exchange rates between sterling and the single currency for an appropriate period of time, there will be few economic obstacles in the way of the UK becoming a member of EMU very early in the twenty-first century.

On 30 March 1998, less than a week after the Commission had confirmed the initial membership of EMU, the Council of Ministers of the EU met in Brussels to begin accession talks with their counterparts from five Central and East European candidate states (the Czech Republic, Estonia, Hungary, Poland, and Slovenia), together with Cyprus. From the outset it was clear that new members could not expect a British-style opt out from the single currency, nor, prior to membership, expect to receive EU funds (of up to 3 billion ECU) to develop transport links and other infrastructure facilities, modernise agriculture and assist business development, unless the relevant candidates closed down their Soviet-era nuclear power stations, put VAT systems in place, granted full citizenship rights to ethnic minorities and set up fully independent judiciaries.

Clearly, because the EMU will impose fiscal and monetary constraints on the macro-economic policy of member countries, the extent to which regional aid can be funded will be strictly limited. With the enlargement of the EU, funds will be spread more thinly, and many of the poorer regions in Western Europe (several with large agricultural sectors) will cease to be eligible because aid will be diverted to even-poorer members in Central and Eastern Europe (with even larger agricultural sectors) (Table 1.1). However, in determining the distribution of regional aid for the period 2000–06, Ministers, first and foremost, will need to focus their attention on economic inequalities within the existing Union – most notably in respect of the substantial disparities in GDP per capita and rates of unemployment.

Regional disparities in living standards and unemployment

Within the EU, many peripheral areas were economically underdeveloped and were lagging behind the rest of the Union in terms of both GDP per capita and employment. There is evidence that peripherality undoubtedly has an adverse effect on 'the levels of innovation, new firm formation and the extent of external control between regional economies' (Tomkins and

Table 1.1 Gross domestic product per capita and employment, EU and Central and Eastern European countries, mid-1990s

	GDP per capita 1996 (purchasing power standard)	% agricultural employment 1996
Luxembourg	169 (1994)	3.8 (1995)
Belgium	114 (1994)	2.7 (1995)
Denmark	116	4.4
Germany	110	3.2
France	107	4.9
Netherlands	105 (1994)	3.7 (1994)
Italy	104	7.5
United Kingdom	101	2.1
EU average	100	5.3
Sweden	98 (1994)	3.3
Ireland	88 (1994)	12.0
Spain	77	9.3
Portugal	68	11.5
Greece	66	20.4
Slovenia	59	6.0
Czech Republic	57	11.0
Hungary	37	8.0
Poland	31	26.9
Estonia	22	14.0

Source: Eurostat and OECD

Twomey, 1994: 157). Broadly, Greece, southern Italy (including Sardegna and Sicilia), Corse, Spain, Portugal, Ireland and the Highlands and Islands of Scotland were all areas so disadvantaged. All had a disproportionately large agricultural working population and a smaller than average workforce employed in industry. Other areas where development was lagging behind included the Eastern *Länder* of Germany which, with their old industrial base and small service sector, contained a concentration of industrial and comparatively unskilled labour. There are also a number of disadvantaged areas which, although are non-peripheral, have rates of unemployment and industrial employment higher than the EU average and where industrial jobs are in structural decline. A large number of these areas exist within the coalfield regions of North-West Europe, for example, Central Scotland, the north-east of England, West Cumbria, South Yorkshire, the north-west of England, South Wales, the Nord-Pas-de-Calais and Est regions of France, parts of the Région Wallonne in Belgium, and Nordrhein-Westfalen in Germany. Apart from a decrease in coalmining, other staple industries were also in decline in these areas, such as iron and steel manufacture, shipbuilding and textiles. Each industry faced changing patterns of competition, demand and output, and since the 1960s coalfields in Western Europe as a whole 'have lost 57 per cent of their production and 82 per cent of their workforce, while the iron and steel industry has lost 60 per cent of its workers' (Tomkins and Twomey, 1994: 157). Away from the coalfield regions, areas of industrial decline can also be found in, for example, sub-regions centred on Barcelona, Torino, Wien and in parts of Central and North-East Sweden. Throughout the EU, there are also large numbers of disadvantaged and vulnerable rural areas with a low level of socio-economic development. In, for example, large parts of western and southern France, north-eastern Spain, central Italy, northern Germany, the south-west and north of England, Wales and the borders of Scotland, there is a high proportion of employment in agriculture, a low level of agricultural incomes, and either a low population density or a high degree of out-migration.

In contrast, there are many regions within the core of EU where economic development is proceeding at a rapid pace and where the average level of prosperity is high. Based disproportionately on high-tech and service employ-ment, and research and development activity, these areas include, for exam-ple, the contiguous *Länder* of Hessen, Bayern and Baden-Württemburg, and the neighbouring *regioni* of Lombardia and Emilia-Romagna, together with the non-contiguous regions of Hamburg, Bremen, Östosterreich, the Ile de France and Région Wallonne.

Table 1.2 and Figure 1.1 show the extent to which living standards are disparate throughout the EU. Whereas, the GDP per capita of the most prosperous region, Hamburg, was 96 per cent higher than the average for the EU in 1994, in the least prosperous region, the Açores, the GDP per capita was as low as 48 per cent of the EU average. Taking the average for the top ten regions, the GDP per capita was as high as 53 per cent above the EU average, but as little as 58 per cent of the EU average for the lowest ten regions. Disparities in rates of unemployment were similarly substantial. Table 1.3 and Figure 1.2 indicate that in 1995 unemployment ranged from as little as 2.7 per cent in Luxembourg to as much as 31.8 per cent in Sur. Taking the lowest ten regions, the average unemployment rate was as low as 5.5 per cent, but in the top ten regions was as high as 22.7 per cent. The spatial distribution of long-term unemployment, moreover, very broadly overlaps the pattern of overall unemployment. Taking the ten regions with the smallest amounts of overall unemployment, the proportion of long-term unemployed averaged only 45.4 per cent of the total, whereas within the ten regions with the largest amounts of overall unemployment, the proportion that was long term amounted to as much as 58.9 per cent.

Clearly within the Member States of the EU, regional disparities in GDP per capita and unemployment are not so marked as in the Union as a whole. There are, however, substantial differences in degrees of disparity between countries (Figures 1.3 and 1.4). Whereas in 1994 the GDP per capita in Germany ranged from 96 per cent above the EU average in Hamburg to only 57 per cent of the EU average in Mecklenburg-Vorpommern (a 109 percentage points difference), in Greece the GDP per capita ranged from only 73 per cent to 57 per cent of the EU average (a 16 percentage points

Table 1.2 Gross domestic product per capita, regions of the EU, 1994

Region	Gross domestic product per capita (purchasing power standard) (EUR 15 = 100) 1994
Hamburg (D)	196
Région Wallonne (B)	183
Luxembourg (L)	169
Ile de France (F)	161
Bremen (D)	156
Hessen (D)	152
Lombardia (I)	131
Bayern (D)	128
Emilia-Romagna (I)	128
Ahvenanmaa/Ahland (SF)	126
Baden-Württemberg (D)	126
Östosterreich (A)	122
Lazio (I)	119
Nord Est (I)	119
South East (UK)	117
Nord Ovest (I)	116
Bruxelles (B)	115
Denmark (DK)	114
West-Nederland (NL)	113
Nordrhein-Westfalin (D)	112
Westosterreich (A)	110
Centro (I)	107
Saarland (D)	106
Schlesewig-Holstein (D)	106
Niedersachsen (D)	105
Berlin (D)	104
Centre-Est (F)	102
Noord-Nederland (NL)	102
Zuid-Nederland (NL)	101
East Anglia (UK)	100
Est (F)	100
Rheinland-Pfalz (D)	100
Bassin Parisien (F)	98
Scotland (UK)	98
Sweden (S)	98
Madrid (E)	95
South West (UK)	95
East Midlands (UK)	93
Oost-Nederland (NL)	93
Sud-Ouest (F)	93
Manner-Suomi (SF)	91
Meditérranée (F)	91
Ouest (F)	91
Vlaams Gewest (B)	91
West Midlands (UK)	90
Noreste (E)	89
Ireland (IRL)	88

North West (UK)	88
Abruzzi-Molise (I)	87
Nord-Pas-de-Calais (F)	87
Sudosterreich (A)	87
Yorkshire & Humberside (UK)	87
Este (E)	86
North (UK)	85
Wales (UK)	81
Northern Ireland (UK)	80
Sardegna (I)	78
Canarias (E)	75
Attiki (GR)	73
Sicilia (I)	70
Campania (I)	69
Continentale (P)	68
Sud (I)	68
Nisia Aigaiou, Kriti (GR)	67
Centro (E)	65
Brandenburg (D)	64
Noroeste (E)	64
Voreia Ellade (GR)	62
Sachsen (D)	60
Sachsen-Anhalt (D)	60
Thüringen (D)	60
Sur (E)	58
Kentriki Ellade (GR)	57
Mecklenburg-Vorpommern (D)	57
Madeira (P)	52
Açores (P)	48

Source: Office for National Statistics (1997), *Regional Trends 32*, The Stationery Office, London

difference). Rates of unemployment also varied substantially between countries. Whereas in 1995 the rate of unemployment ranged from 6.0 per cent of the working population in the Nord Est region to 25.9 per cent in Abruzzi-Molise (a difference of 19.9 per cent), in the Netherlands unemployment ranged from 6.9 per cent in Zuid-Nederland to only 8.9 per cent in Noord Nederland.

It is widely accepted that, both within individual countries and the EU as a whole, marked regional disparities in living standards and unemployment are economically disadvantageous. In areas where GDP per capita is high and unemployment is low, it is probable that economic growth will be at a comparatively rapid rate, there will be a consequential shortage of labour, land and capital in relation to demand and there will be a tendency for cost-push inflation or demand-pull inflation or both to be endemic. Conversely, in areas where GDP per capita is low and unemployment is high, it is probable that the rate of economic growth will be comparatively slow and that labour, land and capital will be under-utilised. However, with some regions being rich and some being poor, and with some suffering from high

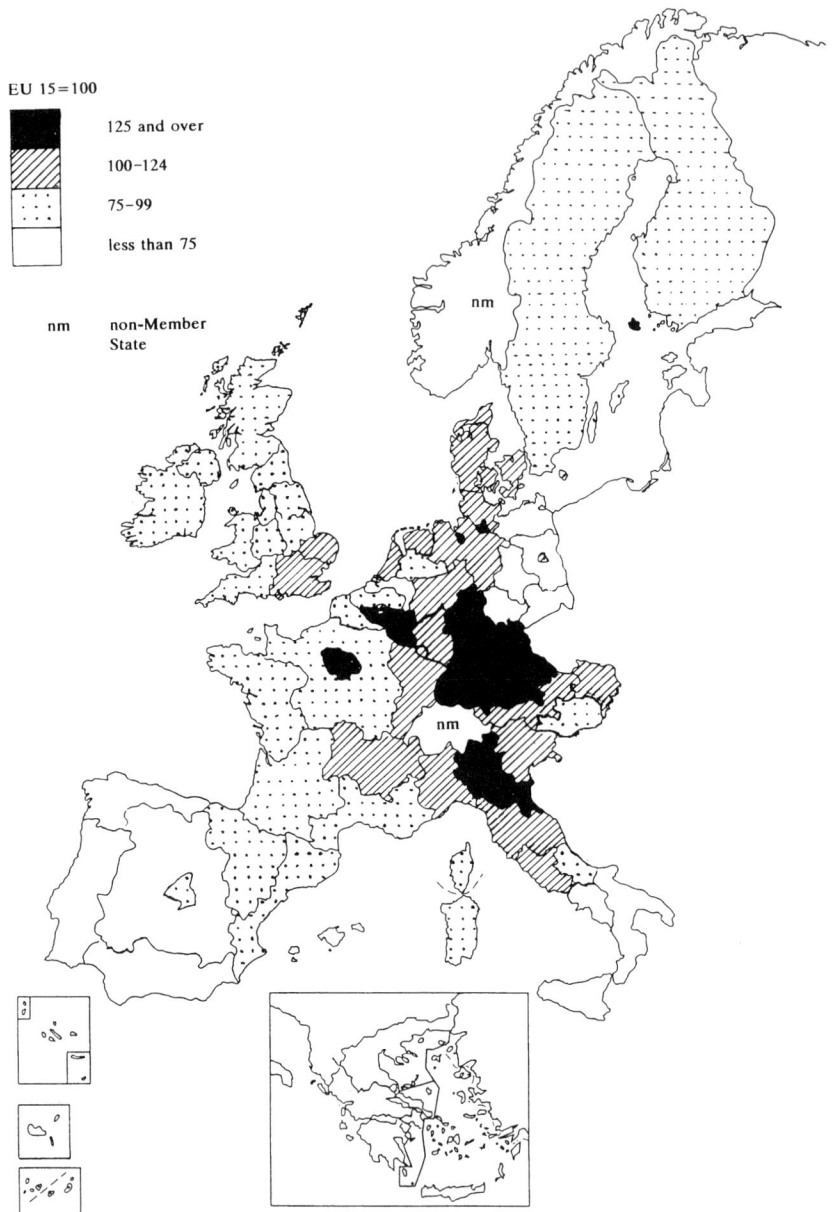

Figure 1.1 Gross domestic product per capita, EU, 1994

Table 1.3 Unemployment rates, regions of the European Union, 1995

Region	Unemployment rate (%), 1995	Long-term unemployed as a % of total unemployment, 1995
EUR 15	10.7	n.a.
Luxembourg (L)	2.7	24.0
Nisia Aigaiou, Kriti (GR)	4.5	50.7
Madeira (P)	4.6	67.2
Bayern (D)	4.9	39.6
Baden-Württemberg (D)	5.5	44.3
Nord Est (I)	6.0	43.8
Lombardia (I)	6.1	50.1
Ahvenanmaa/Aland (SF)	6.2	n.a.
Rheinland-Pfalz (D)	6.2	45.5
Emilia-Romagna (I)	6.3	35.1
Hessen (D)	6.3	51.4
Schleswig-Holstein (D)	6.5	52.6
East Anglia (UK)	6.7	35.2
Zuid-Nederland (NL)	6.9	46.4
Continente (P)	7.1	48.1
Denmark (DK)	7.1	28.2
Oost-Nederland (NL)	7.1	42.9
West-Nederland (NL)	7.3	43.4
Kentriki Ellade (GR)	7.4	63.3
Hamburg (D)	7.6	45.9
South West (UK)	7.6	43.9
Açores (P)	7.8	60.0
East Midlands (UK)	7.8	32.5
Niedersachsen (D)	7.9	46.2
Centro (I)	8.0	60.4
Nordrhein-Westfalin (D)	8.2	37.1
South East (UK)	8.6	44.1
Est (F)	8.7	40.5
Nord Ovest (I)	8.7	63.2
Wales (UK)	8.7	37.3
Noord-Nederland (NL)	8.8	47.4
Scotland (UK)	8.8	40.5
West Midlands (UK)	8.8	51.3
Saarland (D)	9.1	39.1
Sweden (S)	9.1	n.a.
Yorkshire & Humberside (UK)	9.1	38.3
North West (UK)	9.2	45.8
Voreia Ellade (GR)	9.2	45.7
Vlaams Gewest (B)	9.4	61.8
Ile de France (F)	10.0	45.7
Centre-Est (F)	10.2	43.2
Ouest (F)	10.5	33.0
Bremen (D)	10.6	44.7
Campania (I)	10.8	61.6

Table 1.3 continued

Region	Unemployment rate (%), 1995	Long-term unemployed as a % of total unemployment, 1995
Attiki (GR)	11.0	50.2
North (UK)	11.0	44.3
Sud-Ouest (F)	11.0	40.9
Berlin (D)	11.2	53.2
Bassin Parisien (F)	11.9	38.0
Mecklenburg-Vopommern (D)	11.9	60.9
Thüringen (D)	11.9	63.2
Lazio (I)	12.8	62.4
Région Wallonne (B)	12.9	71.3
Northern Ireland (UK)	13.0	51.6
Bruxelles (B)	13.3	81.5
Sachsen (D)	13.8	58.5
Ireland (IRL)	14.3	51.1
Mediterranee (F)	14.4	48.0
Brandenburg (D)	15.1	51.3
Nord-Pas-de-Calais (F)	15.3	53.7
Sachsen-Anhalt (D)	16.7	56.2
Manner-Suomi (SF)	18.2	30.0
Noroeste (E)	18.5	63.0
Sud (I)	18.8	62.8
Noreste (E)	19.3	58.6
Este (E)	20.3	53.0
Madrid (E)	20.7	62.1
Sardegna (I)	20.8	62.4
Centro (E)	22.4	49.3
Sicilia (I)	23.3	65.9
Canarias (E)	23.7	51.7
Abruzzi-Molise (I)	25.9	72.8
Sur (E)	31.8	51.0

Source: Office for National Statistics (1997) *Regional Trends 32*, The Stationery Office, London

unemployment while others have low unemployment, the application of corrective macro-economic policy at times of economic instability can easily have perverse effects. If the national or EU economy is suffering from recession, the introduction of a battery of monetary and fiscal measures to reflate the level of aggregate demand, while benefiting the poorer regions, might very easily over-heat areas of high GDP per capita and low unemployment. Conversely, when inflation is endemic, the use of monetary and fiscal policy to disinflate aggregate demand might very quickly exacerbate the problems of areas with a low GDP per capita and high unemployment. Clearly, a more spatially balanced pattern of living

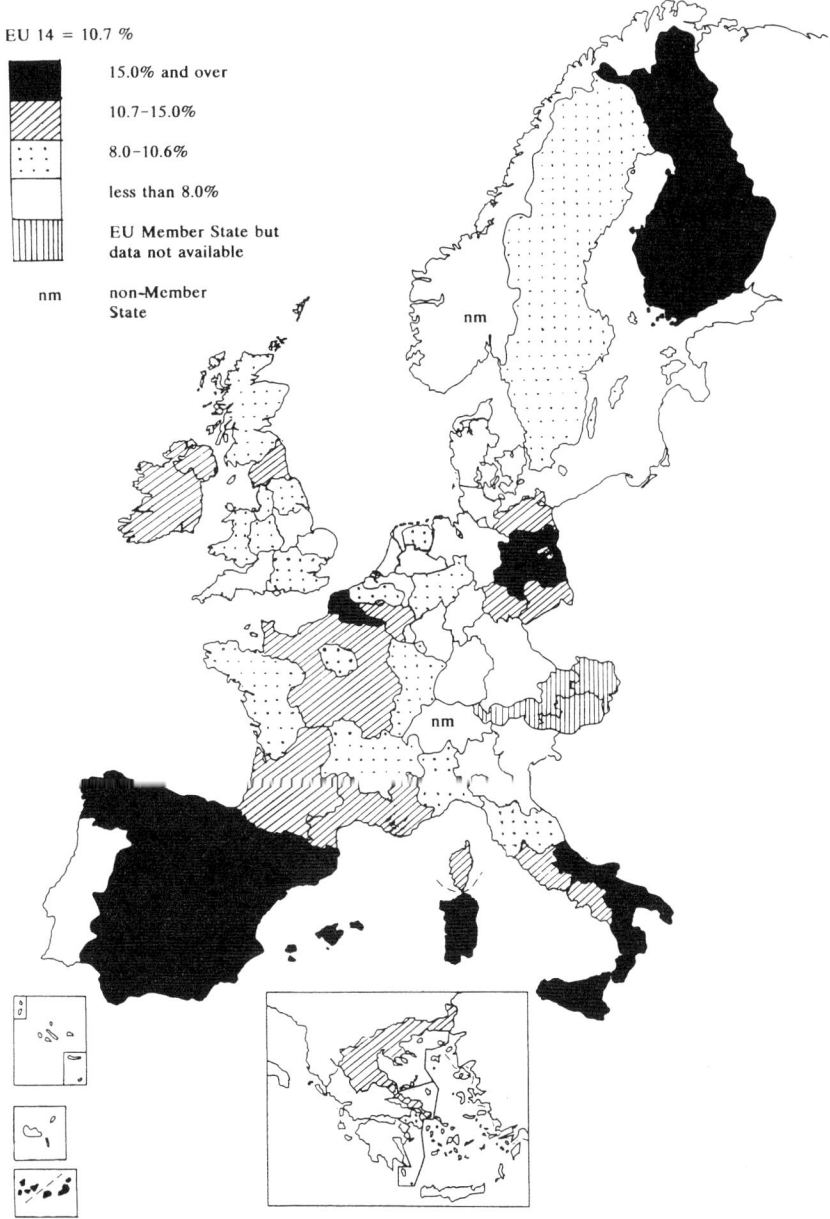

EU 14 = 10.7 %

■ 15.0% and over

▨ 10.7–15.0%

⋮ 8.0–10.6%

☐ less than 8.0%

▥ EU Member State but
data not available

nm non-Member
State

Figure 1.2 Unemployment, EU, 1995

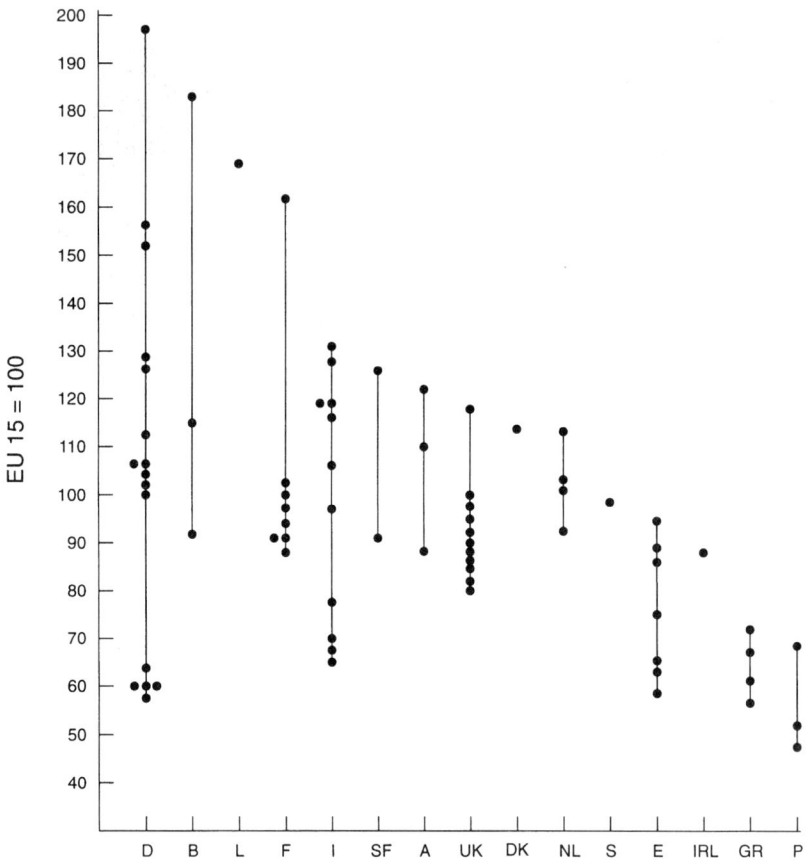

Figure 1.3 National disparities in gross domestic product per capita, 1994

standards and unemployment levels would enhance the efficacy of macro-economic policy when applied to the task of countering inflation or recession on a national or EU scale, and arguably the narrowing of economic inequalities would provide a means of promoting social harmony and political good-will between the regions and Member States of the EU.

Structure and content

Although there are many excellent books in print both on the geography of the European Union and on urban and regional planning in Europe (particularly J. Cole and F. Cole's *A Geography of the European Union*, P. Newman and A. Thornley's *Urban Planning In Europe*, and R.H. Williams' *European Union and Spatial Policy and Planning*), there is an absence of a wide-ranging text covering regional policy both nationally throughout most of Western

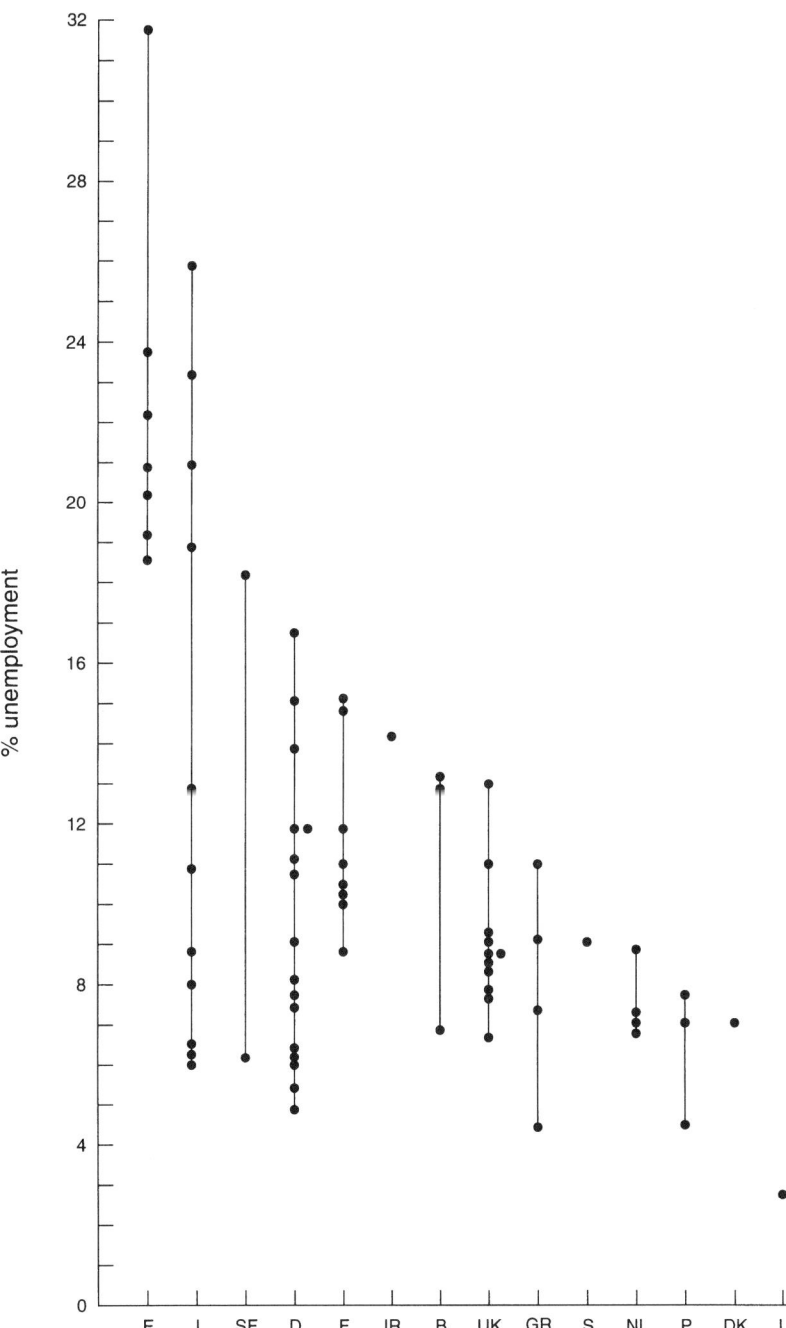

Figure 1.4 National disparities in unemployment, 1995

and Central Europe and within the policy framework of planning on a Europe-wide scale. It is particularly within the context of disparities in living standards and unemployment that Chapter 2 seeks to explain the economics of integration in Europe and provides an overview of the development of institutions and policy within the European Union. Chapter 3 is an analysis of the attributes of the different systems of government and planning which exist in Europe and considers the degree to which regional government might be emerging throughout much of the Continent. In more detail, Chapters 4 to 8 set out to explore the degree to which regional planning and policy has emerged in the different countries of Western Europe – from the highly centralised states of Greece, Ireland and Luxembourg to the Federal states of Germany, Austria and Belgium. Chapter 9 is a detailed analytical study of regional planning and policy in some of the transitional states of East Central Europe. Chapter 10 deals with infrastructure and the environment. Chapter 11 focuses on urban problems and policies, and Chapter 12 concludes by focusing on cohesion and suggesting how the formation of EMU and the enlargement of the EU will affect planning and policy. The issue of rural planning and agriculture is deliberately omitted from the book since, with the reform of the common agricultural policy, relevant subject matter is outside of the scope of this publication and worthy of separate analysis elsewhere.

We must acknowledge a very great debt we owe to present and past colleagues and to a wide range of economists, geographers and practising planners who have stimulated and advised us in the preparation of the book. In respect of Chapter 9, we would particularly like to thank Jiri Blazek and Radim Perlín for their helpful advice on regional policy and planning in the Czech Republic, Gyorgyi Barta and Zoltán Kovács for their assistance with Hungary, and Anna Karwinska for her help on Poland. With regard to Chapter 10, we are grateful to Joseph Watson of the University of Greenwich for contributing a section on the environment, and for compiling the list of references. In addition, we are indebted to Pauline Newell who painstakingly processed much of the manuscript and to Sue Lee and Angela Alwight who produced some of the artwork within the text. Last, but not least, we would like to thank our respective families for their continual encouragement and patience.

Paul Balchin, Luděk Sýkora, and Gregory Bull, Spring 1998

2 The economic basis of integration

Given that the Rome Treaties setting up the European Economic Community had been signed in 1957 and the European Customs Union removing tariffs on trade between Member States was completed in 1968, it is perhaps strange that agreement to deal with regional disparities had to wait until the Paris Summit in 1974 and the consequential introduction of the European Regional Development Fund (ERDF) in 1975.

In the intervening period, The Hague Summit (1969) had agreed to open negotiations for the entry of new members (Britain, Denmark and Ireland subsequently joined in 1973). Moreover, further steps were agreed to provide the Community with its own budget, and to increase the powers of the European Parliament. Economic and Monetary Union (EMU) became an official objective of the Community and the Werner Plan, with the optimistic aim of achieving this by 1980, was in principle agreed by the Member States in 1972.

An agreement to restrict currency fluctuations between Member States (known as the 'Snake') to ± 2.25 per cent was introduced in March 1972, but given the turmoil in currency markets following the 1974 oil crisis, by 1977 only Germany, Benelux and Denmark remained.

Initially, and for some time, ERDF spending was quite small and widely dispersed. From 1975–84, 11.7 billion ECU (£7 billion) was allocated, over 80 per cent going towards infrastructure projects (mainly transport, energy and water) and by the mid-1980s ERDF spending represented only 7.3 per cent of the Community's budget (Balchin and Bull, 1987). This compares with a figure of 17 per cent in 1989 and 26 per cent in 1992 on a much larger budget.

The initial fund was, however, much smaller than the one proposed by the Commission. The Report on the Regional Problems in the Enlarged Community (The Thomson Report) had stated that, whilst the ERDF should be used to generate growth in less developed and declining regions, a common regional policy would not be a substitute for national regional policies, but rather a complement to them. The fund operated by partially reimbursing the expenditure of Member States on infrastructure works and investment incentives in the assisted regions.

Although no specific mention of regional policy was included in the Treaty of Rome, the Preamble nevertheless refers to 'reducing the differences existing between . . . regions and . . . mitigating the backwardness of the less favoured'. In practice, the reasons for establishing the ERDF were more pragmatic. First, enlargement would include new members with severe regional problems (Ireland and Britain). The regional situation in these countries – as elsewhere – was soon aggravated by the recessionary effects of the 1974 oil crisis. Second, it was acknowledged that the process of integration and enhanced competition might itself lead to a worsening of the regional situation in some areas. The ERDF has been continuously enlarged and revised since 1975 (notably in 1984–85 and 1989 and 1992). Under the earlier changes requests for funding had to conform to certain criteria and would no longer be handed out automatically. The common regional policy would include analysis of the socio-economic situation of the regions, co-ordination of national regional policies and an assessment of the regional impact of major policies.

The 1988 reforms identified three regional development 'objectives' (in favour of lagging regions, areas with industrial decline and rural areas) to be tackled at the Community level by the structural funds, including the ERDF, and other financial instruments, thus rationalising fund intervention. In addition, the 1988 reforms introduced the aim of concentrating funding on a limited number of priorities within a region's 'Community support framework', bearing in mind the specific needs of each region.

In accordance with the Brussels' budgetary agreement of 1988, Structural Funds (which included not only the ERDF but also the European Social Fund and the Guidance section of the EAGGF) were increased from ECU 58.4 billion, 1958–88, to ECU 64 billion (at 1989 prices) for the period 1989–93.

Following revisions in 1993, budgetary allocations to the Structural Fund were increased again to account for nearly a third of the budget, or ECU 154.5 billion (at 1994 prices) for the period 1994–99. Almost 61 per cent of this spending would be concentrated in the Objective 1 areas (the lagging regions) and only 10 per cent in Objective 2 areas (areas of industrial decline) and 4.5 per cent in Objective 5b areas (vulnerable rural areas) (Table 2.1). However, as Figure 2.1 shows, these shares would vary considerably from country to country and, as in previous periods, the funds were, to a large extent, subject to national quotas by objective. Such concentration of spending on lagging regions (with GDP below 75 per cent of the EU average) has greatly concentrated resources in weaker Member States and those with severe regional problems: overall Spain would receive 22.5 per cent, Italy and Germany 14 per cent and Greece and Portugal 10 per cent each of Structural Fund spending from 1994–99. With continuous enlargement to fifteen members of the Community (Greece joined in 1981 and Spain and Portugal in 1986), Britain could

Table 2.1 Structural assistance by country and objective, 1994–99

Country	Objective 1	Objective 2	Objective 3 & 4	Objective 5a	Objective 5b	Objective 6	CIs	Total	%
			(ECU million at 1994 prices)						
Spain	26,300	2,416	1,843	446	664	-	2,774	34,443	22.5
Germany	13,640	1,566	1,942	1,143	1,227	-	2,206	21,724	14.2
Italy	14,860	1,463	1,715	814	901	-	1,893	21,646	14.1
Greece	13,980	-	-	-	-	-	1,151	15,131	9.9
Portugal	13,980	-	-	-	-	-	1,058	15,038	9.8
France	2,190	3,774	3,203	1,933	2,238	-	1,601	14,938	9.8
UK	2,360	4,581	3,377	430	817	-	1,570	13,155	8.6
Ireland	5,620	-	-	-	-	-	483	6,103	4.0
Netherlands	-	650	1,079	165	150	-	421	2,615	1.7
Belgium	730	342	465	195	77	-	287	2,096	1.4
Finland	-	179	336	347	190	450	150	1,652	1.0
Sweden	-	157	509	204	135	247	125	1,377	0.9
Luxembourg	-	15	23	40	6	-	20	104	0.1
Total	93,972	15,360	15,180	6,918	6,862	697	14,051	153,038	100.0
%	61.4	10.0	9.9	4.5	4.5	0.5	9.2	100	

Source: CEC (1996a) *Europe at the Service of Regional Development*

Notes
Objective 1   The economic adjustment of regions whose development is lagging behind.
Objective 2   The economic conversion of declining industrial areas.
Objective 3   Combating long-term unemployment, etc.
Objective 4   The adaptation of workers to changes in industry, etc.
Objective 5a Adapting structures in agriculture and fisheries.
Objective 5b The economic diversification of vulnerable rural areas.
Objective 6   Development of regions with an extremely low population density.
CIs   Community Initiatives.

Only Objective 1, 2, 5b and 6 assistance, and CIs are regionally-targeted.

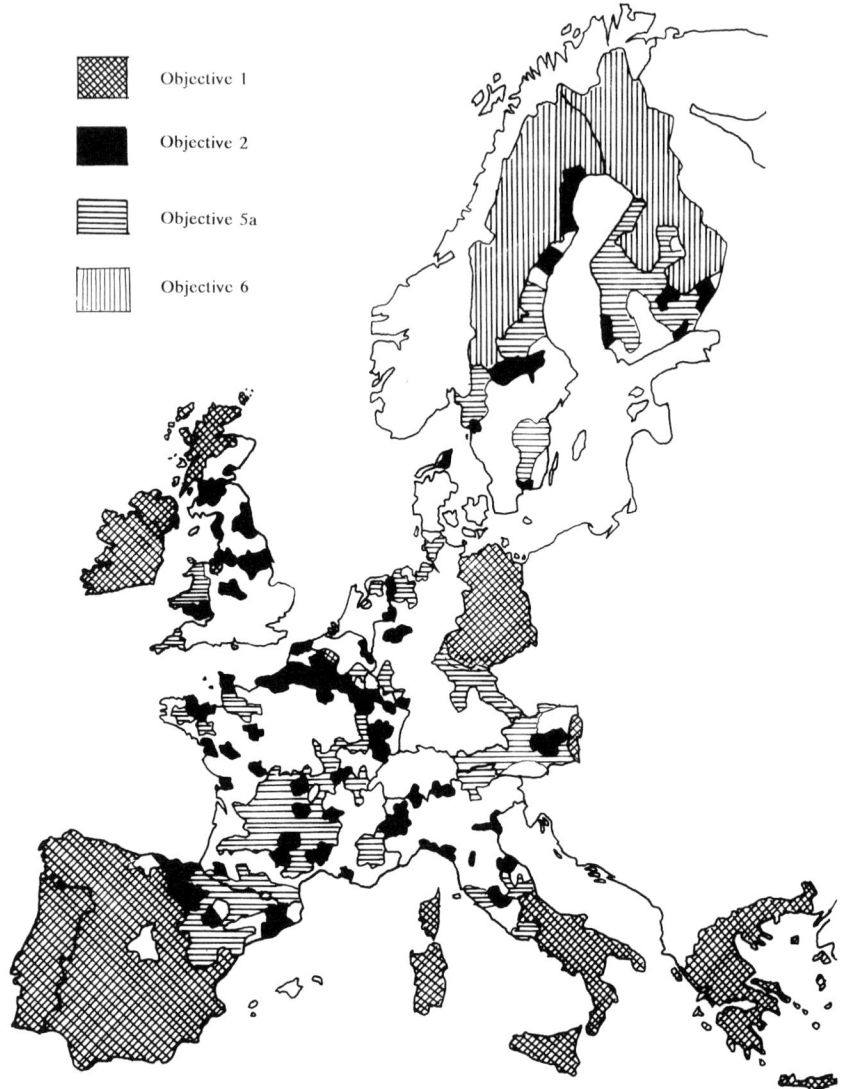

Figure 2.1 Designated Structural Fund areas, EU, 1995

no longer expect to maintain its share of ERDF spending, which had reached 24 per cent over the period 1975–85 (Bulletin EC, 12: 1985). In turn this would have financial implications for the UK which, as a significant net contributor to the EC budget (largely through the operation of the CAP), would no longer be able to rely on Structural Fund allocations to reduce significantly its overall net contributions.

Negotiations on revisions to the Structural Fund in the early 1990s took place against the background of the European Council meeting in Maastricht in 1991. The protocol to the Maastricht Treaty on European Union recognised that measures to achieve economic convergence towards EMU would be endangered without associated action to improve economic and social cohesion. Mainly in view of the financial and budgetary implications of the Commission proposals, known as the 'Delors II package', opinion remained divided over the proposals to raise spending on structural operations considerably. Eventually a consensus was reached at the Edinburgh Summit in 1992. In addition to an agreement to increase Structural Fund spending to 154.5 billion ECU for the period 1994–99, a Cohesion Fund budget benefiting the weakest Member States (Portugal, Greece, Ireland, Spain) was set at 15.15 billion ECU for the same five years. Together these changes (at 1992 prices) represented a doubling of receipts by Greece, Portugal, Ireland and Spain compared to 1992, and an almost four-fold expansion compared to 1988 (CEC, 1993a). The revised Regulations adopted in 1993 (OJEC No. L193, 31.7.93) involved relatively minor changes to the operation of the Funds. A slightly higher concentration of spending (61.4 per cent) was to occur in Objective 1 areas. There would be a greater emphasis on prior appraisal of projects and ex post-evaluation, and a new Structural Fund (FIFG) would help support diversification in the Fisheries sector. In administration, there would be a reduction in the number of planning stages from three to two. In most cases, instead of the three-stage process of Regional Development Plans (RDP), Community Support Frameworks (CSF) and Operational Programmes, regions can now prepare Single Programming Documents (SPDs) containing their overall development plan as well as the operational detail.

In terms of coverage, the five new German *Länder* and East Berlin were included under Objective 1, covering nearly 21 per cent of the German population. In addition, certain parts of the richer EU countries (e.g. Merseyside, Highlands and Islands and Nord-Pas-de-Calais) were included under Objective 1 even though their per capita GDP was slightly above the original criteria for 'lagging' regions of 75 per cent of the EU average. In total, Objective 1 areas contained 25 per cent of the EU population in 1993, Objective 2 areas 16.4 per cent, Objective 5b areas 8.8 per cent and Objective 6 areas 0.4 per cent (Table 2.2).

In 1995, Sweden, Austria and Finland joined the EU, representing only a 7.4 per cent expansion of the total population. Under the Treaty of Accession additional funds of just under ECU 4.6 billion, a 4.5 per cent increase in the Structural Fund was provided. Objective 6, to promote development of regions with low population density, was introduced for the Scandinavian countries, although spending on this Objective (up to 1999) accounted for less than a fifth of the additional funds.

In addition to Objective 1–6 assistance (which acts in response to requests

Table 2.2 Population of areas eligible for structural assistance under regional objectives, 1993

Country	Objective 1	(% Total Population) Objective 2	Objective 5b	Objective 6	Total
Greece	100.00	-	-	-	100.00
Ireland	100.00	-	-	-	100.00
Portugal	100.00	-	-	-	100.00
Spain	58.20	20.30	4.40	-	82.90
Italy	36.60	10.80	8.40	-	55.80
Finland	-	15.50	21.50	16.60	53.60
France	4.40	25.90	17.30	-	47.60
Luxembourg	-	34.20	7.80	-	42.60
UK	6.00	31.00	4.90	-	41.90
Austria	3.50	8.20	28.90	-	40.60
Germany	20.70	8.80	9.60	-	39.10
Belgium	12.80	14.00	4.50	-	31.30
Sweden	-	11.00	8.60	5.00	24.60
Netherlands	1.45	17.30	5.40	-	24.15
Denmark	-	8.30	7.00	-	15.80
EU	25.00	16.40	8.80	0.40	50.60

Source: CEC (1996b) *Europe at the service of regional development*

for funding from EU members), the Maastricht Treaty incorporated Community Initiatives (CIs) into the process of structural funding. CIs enable the European Commission to take the initiative in promoting regional development, are intended to be transactional and inter-regional, and provide the EU with an opportunity to take a high profile in the selected localities within Objective 1, 2 and 5b areas. Since 1993, CIs have been based on the following seven priorities:

1   cross-border, trans-national and inter-regional co-operation and networking;
2   rural development;
3   development of the most remote regions.
4   employment and development of human resources;
5   management of industrial change;
6   development of crisis-hit urban areas;
7   restructuring of the fishing industry.

In 1994, the Commission adopted no fewer than 13 CIs – amounting in value to ECU 14 billion or 9.2 per cent of structural funding over the period 1994–99 (see Table 2.3).

The Maastricht Treaty also established the Committee of Regions (CoR) in 1994 to give a voice to regional and local authorities in EU debates and

Table 2.3 Community Initiatives, 1994–99

Initiative	Purpose	Ecu million
INTERREG	Cross-border co-operation, the development of energy networks, international co-operation in water management	3,447.6
LEADER	Local development in rural areas	1,722.6
REGIS	Support for the most remote regions: French overseas départements, the Azores and Madeira (Portugal) and the Canaries (Spain)	600.0
Employment (Now HORIZON YOUTHSTART)	Vocation integration of women, young people, the disabled and the excluded	1,784.0
ADAPT	Updating for labour force	1,623.0
RECHAR	Conversion of coal mining areas	448.8
RESIDER	Conversion of steel areas	564.2
RETEX	Diversification of areas dependent on the textile industry	562.8
KONVER	Diversification of areas dependent upon military activities	744.3
SMEs	Improving the international competitiveness of small and medium size firms	1054.0
URBAN	Restoration of crisis-hit areas	800.1
PESCA	Restructuring of the fisheries industry	290.3
PEACE	Support for the process of peace and reconciliation in Northern Ireland and the six border counties of Ireland	300.0

Source: CEC (1996) *Europe at the service of refined development*

policy-making. With its 222 members and eight commissions, the CoR is particularly concerned with the need for greater legislative subsidiarity as a means of strengthening the role of regional and local government. However, although the CoR 'is an attempt to address both the democratic deficit and the implementation deficit' (Williams, 1996: 43), the outcome of its deliberations will inevitably result in greater competition for structural funding among the regions eligible for Objective 1, 2 and 5b assistance.

Future enlargement of the EU is likely to cause even more problems for the Structural Fund. It is possible that the EU could include twenty-five or more members early in the new century. Applications have come from the ten associated Central and Eastern European (CEE) countries, Bulgaria, the Czech Republic, Hungary, Poland, Romania, Slovakia, Slovenia, Estonia, Latvia and Lithuania. These countries have signed Europe Agreements covering free trade in industrial goods, approximation of legislation and economic, financial and political co-operation. The Pre-accession Strategy for these countries involves reforms and the establishment of structures to enable them to meet the obligations of membership (including single market legislation), as well as assistance and transfer of technical know-how, mainly through the PHARE programme, supporting economic re-construction and democratic reform (see Chapter 13).

With a combined population of 106 million, the CEE countries would represent a 29 per cent increase of the EU's population, but less than a 4 per cent increase in GDP. Agriculture accounts for 26.7 per cent of employment in the CEE countries compared to less than 5.7 per cent in the EU, and considerable effort to modernise agriculture and implement reforms would probably be expected by the EU which would, in any case, need to extend price support through the Common Agricultural Policy (CAP) at an estimated cost of 40 billion ECU (*The European*, 1997). Farm production in existing EU Member States would clearly also be affected.

However, given the failure of the 1997 Amsterdam Summit to find agreement on significant internal institutional reform, support appears to be mounting in the European Parliament for a block on any further enlargement of the EU until further reforms are undertaken to enable a much larger union to operate efficiently (EP, 1997).

Although accession negotiations are programmed to open with the Czech Republic, Estonia, Hungary, Poland and Slovenia, it seems probable that, given the lack of real reform, at best only three or four newcomers will be admitted in the next few years. A further intergovernmental conference to examine the proposals for institutional reform for enlargement, is not scheduled until after the year 2000.

Economic integration and the regions

As mentioned earlier, progress towards EMU and the increasing emphasis on economic and social cohesion by the Single European Act (1988) and the Maastricht Treaty (1991) have had a significant impact on the size, scope and implementation of EU policy towards the regions. Echoing the Single European Act, Article 130A of the Maastricht Treaty on European Union states:

> The Community shall develop and pursue its actions leading to the strengthening of its economic and social cohesion. In particular, the Community shall aim at reducing disparities between the levels of development of the various regions and the backwardness of the least favoured regions, including rural areas.

In addition to having to deal with existing regional problems and disparities in economic development, the European Community would need to address the problem of a possible widening of such disparities brought about by the continuing process of European integration.

Following a number of years of internal dispute within the Community, concerning a range of issues, including enlargement, growing surpluses and spending on the CAP, Britain's budgetary position and the Community budget and institutional reform, much progress was made at the 1984 Fontainbleau Summit. Britain's agreement to settle its budgetary dispute and to explore further issues through two committees (dealing with free movement of people and institutional affairs) was undoubtedly influenced by French and German ideas to split the EC into two sub-groups (*Europe à deux vitesses*) to allow willing members to forge ahead with EMU (Swann, 1996). In 1985 the latter of these committees recommended the establishment of an intergovernmental conference to negotiate a European Union Treaty. This occurred a few months later in Milan. Liberalisation of the internal market had a high priority since it enjoyed fairly general support from Member States as well as the European Commission. Significant progress on monetary union was unlikely as Germany had ruled it out until capital flows had been liberalised (Swann, 1996). The eventual outcome – the Single European Act (SEA) was formally adopted in 1986. In the preamble to the SEA, there was notably reference to the fact, and a reminder, that heads of state and government had approved the progressive realisation of EMU at the Paris Summit (1972) (Swann, 1996). Recalling this commitment, the European Council at the Hanover Summit in 1988 asked the President of the EC Commission (Jacques Delors) to examine the process and stages by which EMU might be achieved. The Delors Report (1989) proposed a three-stage plan towards EMU. The first stage involving the elimination of all restrictions on capital movements between Member States was begun on 1 July 1990. The next stages were mapped out in more detail in the Maastricht Treaty (1991); the second stage from 1994 saw convergence criteria targets to be achieved for fiscal and monetary variables and the establishment of the European Monetary Institute (as a pre-figuration of the European Central Bank). A decision was to be taken by the end of 1996 on whether enough countries met the convergence criteria, but this was postponed until 1998. The latest date for the final stage, involving introduction of the single currency (subsequently named the Euro) was 1 January 1999. Britain, Denmark and Sweden however, secured an opt-out of the final stage, but may 'opt-in' later (while Greece failed to qualify).

In some ways the Single European Market (SEM) programme could be seen as a 'tidying up' operation designed to achieve the original objectives of the Treaty of Rome and the establishment of a truly unified European market (Williams, 1996). It is also true to say that the SEM was seen by many as a response to the global challenge of the USA and Japan. By the late 1970s and early 1980s there was growing recognition that Europe lagged behind in key high-tech sectors. The European Commission pointed out that thirty-one technological sectors of the future were dominated by the USA and another nine by Japan and only two by Europe (CEC, 1988a). The lack of a large home market and the presence of technical barriers to trade (in the form of differing national standards even within Europe) meant that European firms were often much smaller than their American and Japanese counterparts, making investment and research into the development of new products more difficult and costly. In 1981, 62 per cent of patents granted in Europe were of European origin, but by 1993 this figure had fallen to 49 per cent, with Japan taking 19 per cent and the USA 28 per cent (CEC, 1994a).

As a trading block accounting for 15 per cent of world exports in 1980, the European Community was more dependent on world trade than either the USA or Japan (11 per cent and 6.5 per cent of world exports respectively) and intra-Community trade accounted for a further 19 per cent of world exports. In addition to having to maintain their competitiveness in world markets, the expansion of intra-European trade meant that Member States could no longer afford to overlook the remaining trade barriers – specifically non-tariff barriers (NTBs) – within Europe. From 1971 to 1990 the percentage of the UK's exports to the European Community increased from 29 per cent to 54 per cent. Exports overall represented nearly a fifth of GDP in the UK and over a quarter of GDP in West Germany, Denmark and Portugal in the mid-1980s. In Ireland, the Netherlands and Belgium and Luxembourg this figure reached 50 per cent and more (Eurostat).

The SEM programme was based on Lord Cockfield's White Paper 'Completing the Internal Market' (CEC,1985). This categorised the barriers to completion of a single market into three types: physical, technical and fiscal. The eventual programme contained some 289 legislative proposals, often with differing time-scales, but with the aim of completion by 1 January 1993.

The Single European Act (1986) defines the SEM as 'an area without internal frontiers in which the free market of goods, persons, services and capital is ensured'. To improve decision-making the Act also included provisions for the extension of majority voting in the European Council (although not on fiscal matters, e.g. VAT) and increased powers for the European Parliament; it added new sections or titles on the environment, and on economic and social cohesion, the latter largely in response to pressure from Spain and Portugal who indicated that their condition for acceptance of the Single Act was that there should be a transfer of resources to weaker economies of the European Community (Swann, 1996).

The SEM legislation concentrated on the following areas:

- *Removing physical barriers*
  Border controls were simplified in 1988 by means of a Single Administrative Document for goods crossing internal borders. Even this was

abolished in 1993 and in most cases controls now operate only on goods from outside the European Union. Liberalisation of transport, particularly road haulage, has also been important in this context.

- *Removing fiscal barriers*
  National differences in indirect taxes (VAT and excise duties in particular) make for major obstacles to cross-border trade. Prices would differ considerably from country to country simply due to different rates of VAT. Harmonisation of VAT rates was eventually agreed in 1997. This allows collection to switch over to the original system, rather than being collected by the country where the goods are consumed and zero rating of exports can cease (Swann, 1996).

- *Removal of technical barriers*
  Although trade within the European Community was free of border taxes or tariffs, many non-tariff barriers (NTBs) to trade remained, and were often of a technical nature and resulted from differing national standards, regulations and subsidies. The effect of such NTBs was that trade between Member States was impeded, or goods had to be modified to meet different standards, often resulting in production of similar goods in several EC counties.

Up to 1985 approximately 180 directives had been adopted relating to industrial products and foodstuffs (concerning labelling, packaging, advertising, composition and additives). Since product standardisation would have been lengthy and over regulatory, in order to speed up the process, the principle of mutual recognition of standards was adopted provided that essential requirements (e.g. safety) were met (Swann, 1996). In addition, the setting of essential technical product requirements was given to special European standards bodies and was pursued by defining product categories rather than single products (e.g. Construction Products Directive and the development of the Eurocodes in construction). A 'CE' mark was introduced to show that products conform with essential requirements.

In addition, the SEM programme involved measures to open up financial services (e.g. banking, pensions and insurance) and national financial markets with the adoption in 1988 of a directive liberalising capital movements. The opening up of 'public procurement' markets was also expected to greatly enhance competition as barely 2 per cent of government contracts were estimated to go to non-national firms. European Commission estimates suggested this sector was important and accounted for as much as 15 per cent of the Community economy, but earlier attempts to open up public contracts to non-national firms had achieved little. Existing directives on procurement (e.g. public supply contracts and public works contracting) were tightened up, and public procurement rules were agreed for public utilities (e.g. water, transport, energy and telecommunications) and services during the early 1990s (Swann, 1996). Finally, the Single European Act introduced further measures to enhance free movement of workers (a right originally covered in Article 48 of the Treaty of Rome), regarding rights to unemployment benefit, residence permits and the elimination of qualifying periods for eligibility for employment.

In spite of resistance from the UK government on the matter of border controls, which Mrs Thatcher claimed essential 'to protect our citizens from crime and stop the movement of drugs or terrorists and of illegal immigrants' (Thatcher, 1988), a number of countries signed the Schengen Treaty in 1990 and agreed to move towards the total elimination of border controls for travel between signatory countries and to develop common visa and asylum policy on their external frontiers. Of the European Community countries only the UK, Denmark and Ireland did not sign, and the issue of border controls was still unresolved after the Amsterdam Summit in 1997.

Benefits of the SEM

The precise nature and extent of the effects resulting from the creation of the SEM were explored in a research programme launched by the Commission entitled 'The Cost of a Non-Europe' (CEC, 1988b; Cecchini, 1988).

The studies suggested that SEM benefits would arise from a number of sources.

- Improved trade would lead to price convergence across countries and lower the average price level for consumers and producers.
- Further cost reductions in production would arise due to:
  a economics of scale associated with fewer, larger production units (plants) and enterprises (firms); and
  b greater specialisation of production and trades; and
  c improved access to new technology and innovations, new processes and products.

The overall effect of the SEM was estimated to give the potential for raising Community GDP by between 4.5 and 7 per cent with the net creation of between 1.75 and 5 million jobs (CEC, 1988b).

There were inevitably criticisms of the report, particularly regarding the lack of regional or even national breakdown of the effects and the assumption that labour made redundant from rationalisation would be easily re-employed elsewhere. The latter point is particularly critical as the report forecasts as many as 500,000 job losses in the first year (Cecchini, 1988). The potential benefits of cost reduction are also dependent on competition being maintained in industrial sectors already characterised by oligopolistic market structures and where the emergence of fewer larger firms can be expected to continue to increase through take-overs and mergers. Such activity within the EC trebled between 1986 and 1992. By the early 1990s, 'strengthening of market position' and 'expansion' were cited as the main reasons for merger and acquisition activity by three out of four EC companies (CEC, 1994b).

The Commission identified 100 sectors which it expected would be potentially vulnerable to change in the SEM and in a later study pointed out the distribution of some forty sensitive industries at the national level (CEC, 1988b, 1990a). Three categories of sectors at risk were identified:

1   Rapidly growing high-tech sectors (e.g. computers, telecommunications equipment) which previously were often restricted by high NTBs (e.g. national regulations) but where there is significant scope for expansion at the European level through investment, mergers or joint ventures.
2   Industries characterised by low import penetration due to high NTBs, where there is wide price variability and where considerable rationalisation and restructuring can be expected (e.g. pharmaceuticals).
3   Industries where NTBs are less substantial but where production is likely to shift towards countries able to achieve cost advantages through economies of scale in the expanded single market (e.g. motor vehicles, shipbuilding, chemicals and clothing).

At a national level these industries are often heavily localised in only a few regions and even if a country has as many winners as losers, the effect at a sub-national or regional level could clearly be one of significant gains or losses depending on:

a   the make up of the local economy; and
b   whether the industrial sectors located there are themselves winners or losers at the European level.

South European countries (Spain, Portugal and Greece) may be expected to benefit by having lower labour costs and the potential to extend economies of scale in a fairly wide range of sectors through improved access to wider markets (e.g. clothing and textiles, motor vehicles).

But whilst firms in these countries have better access to northern markets, the reverse is also true – particularly in the context of intensive high-tech industries and financial and business services in which many more centrally located regions specialise. The Commission's own survey of 9,000 businesses in the EC found that firms in the more prosperous regions tended to have a more positive view of the effects of the single market, whereas firms located in 'problem' regions were least clear of the potential impact of the SEM (CEC, 1990a).

The risk for less industrialised southern European states is that they will increasingly specialise in areas that exploit their regional advantages (i.e. cheaper labour costs), particularly in consumer goods (textiles, clothing, food, timber, etc.) where there are relatively weak growth prospects and where competition from developing countries is increasing (Blacksell and Williams, 1994).

Overall, however, it is likely that the regional employment shifts resulting from the impact of the SEM can be ascribed to a much wider range of factors than economies of scale alone. For example, firms previously involved in pan-European operations with country-based plants in most national EC markets may well rationalise production into fewer (possibly more centrally located) Euro-plants supplying the whole single market in any one product line. The resulting benefits may come not only from longer production runs and economies of scale, but also from the reduction of generous buffer capacities often required in smaller country-based plants (Collins and Schmenner, 1995) and from a consequent reduction in stockholding. Against this, total transport costs would almost certainly rise as goods would have to travel farther to market, as might raw materials or components to the factory. The question of whether producers may perceive a single market or several smaller market clusters based on cultural or other considerations, may well influence the extent of rationalisation and the scale of potential gains, and according to Phelps (1997) the existence of market clusters may provide better marketing advantages for more peripheral countries.

Supply-side factors relating to labour skills, transport and communications infrastructure, access to technology and innovative capacity as well as other region-specific characteristics, factor costs and location relative to main markets, will also be important in explaining the performance of existing industries, levels of new investment and inward investment into the regions.

Regional convergence or divergence?

It is important to assess the relative importance of the above-mentioned competitiveness factors in determining the best combination of regional policy measures for particular types of regions (CEC, 1991a). This section provides a brief typology of the different types of regions and examines in more detail whether the costs and benefits of integration are likely to be spread evenly across the regions.

Bearing in mind that regions may fall into more than one category, the main types of problem regions can be described under the following headings.

Industrial conversion regions

These are often, but not exclusively, located in the older, centrally-located industrial regions of Europe. Basic industries such as iron and steel, heavy engineering and extractive industries (including coalmining) tended to predominate over more dynamic manufacturing and high-tech sectors in these regions. Included in this list would be South Wales, Teesside in North-East England, the East Midlands, Glasgow, parts of southern Belgium, northern and eastern France, northern Spain and parts of the Ruhr and Sarre in Germany. In addition, one should also include old textile regions heavily dependent on the industry (e.g. Nord-Pas-de-Calais) and declining port areas (e.g. Liverpool). These sectors have frequently been disadvantaged by a combination of falling demand, rising overseas competition, lack of investment and, in some cases, locational factors (e.g. greater advantage of coastal locations for iron and steel). These areas have had to try to overcome a number of associated problems; the resultant dereliction of land and buildings, the lack of training in skills required for new industries, a weak tradition of local entrepreneurship, poor infrastructure and the need to compete for investment with emerging high-tech regions elsewhere in the same country (e.g. M4 corridor and Cambridge sub-region in UK, Bayern in Germany). The UK has the highest proportion of regions in this category, as illustrated by its high share of Objective 2 structural fund spending. The main aim for these regions is to diversify into other economic sectors. Although by the mid-1980s only a very few regions in Europe could still be classified as being dominated by these sectors (Blacksell and Williams, 1994), this outcome probably has more to do with often massive employment decline in basic industries rather than with substantial diversification and growth elsewhere. Champion *et al.* (1996) show that regions defined as 'early heavy manufacturing', recorded the most marked and consistent fall in regional GDP per capita from 1960–90, falling to just below the Western European average.

Problem urban regions

Like the former, these tend to be concentrated more heavily (but not exclusively) in and around the core central areas of Europe where population density is highest. Such areas provide important agglomeration economies for industry and services, including proximity to other firms and access to a large, well-qualified labour force, and are often endowed with major financial/business service and retail centres. This concentration can in itself cause environmental problems through congestion, noise and atmospheric pollution (CEC, 1994c: 104), especially in cities with extensive commuting hinterlands such as Greater London, the Paris region, Greater Athinai, the Randstad region in the Netherlands (which contains nearly half the national population in just four cities), the Rhine-Ruhr area of Germany and many others. Smaller cities and towns are also widely affected by congestion, particularly where they contain historic centres, older road and transport infrastructure and generally less extensive mass transit systems than larger capital cities.

In addition, the greater concentration of economic activity and population is reflected in land values which peak within the central business district but remain high in outlying areas and often throughout the wider agglomeration. Such values are reflected in house prices and rents, up to double the national average, and although social housing schemes can help provide housing at below market rents, many countries are reducing state involvement and subsidy to social housing. Of course, many groups may be denied access to social housing in the first place, such as young single persons or households whose incomes fall just above qualification levels, and these groups are largely dependent on an increasingly expensive rented sector, which is itself diminishing through gentrification and urban re-development (Bull, 1996). Access to affordable housing often combines with problems of unemployment or low paid employment; a recent study by Eurostat (1997) indicates that 57 million people in the European Union (12) in 1993 (some one in six) were affected by poverty and social exclusion with just over a third of these consisting of the 'working poor', that is people who had jobs but earned below poverty wage levels.

Employment decline in manufacturing has often continued at above average levels in the large cities, even where industrial structure has not been unfavourable; Champion *et al.* (1996) confirm that in Western European 'central metropolitan regions' (i.e. the main capital regions, for example London, Paris, Randstad, Bruxelles, but also including peripheral ones such as Roma, Madrid and Stockholm), de-industrialisation over the 1980s proceeded at a much faster rate than average, so that by 1990 only a fifth of their GDP came from manufacturing. By contrast, these areas saw growth of banking, insurance and market services by more than half as much again as the average and the contribution of these sectors grew by almost a third, coming to represent 37.6 per cent of GDP by 1990 (Champion, *et al.*, 1996).

Employment opportunities have thus become polarised in Europe's major city regions as manufacturing jobs have been replaced by lower paid jobs in personal services, retailing, hotels and catering, distribution and, at the other end of the scale, more highly qualified jobs in the office sector.

Social exclusion is often high due to the strong incidence of marginalised groups such as the young and immigrant populations, often attracted by the size and diversity of the economies of most large cities and the range of jobs available. Between 1987 and 1991, for example, the population of German cities swelled by an average of 18 per cent as two-thirds of the substantial numbers of immigrants from Central and Eastern Europe and the former Soviet Union moved to cities of over 500,000 people.

Finally de-industrialisation has contributed to a legacy of vacancy and dereliction in many cities. Coal and steel districts in the Ruhr, Belgium and France have been seriously affected by closure at various times, and port areas of Merseyside, Hamburg, Rotterdam and Duisburg have been greatly affected by dockland changes. Britain's cities have probably suffered more than elsewhere in Europe – in the late 1980s a DoE study revealed that around 40,500 ha of land was derelict (about half of which was urban) and at roughly the same time the Lacaze Report in France found a total of 20,000 ha of dereliction, the highest proportion of which appeared to be highly localised, notably in the Pas-de-Calais region followed by Lorraine with much smaller proportions in the industrial districts of Paris and in the Rhone Alps (Kivell, 1993).

Policies for these areas have revolved around economic development and the re-use of buildings and land, retraining and the acquisition of new skills and, more recently, on environmental improvements, transport infrastructure (e.g. light rail) and other efforts to reduce social exclusion (community centres, hostel accommodation, cultural, leisure and sports amenities, drug rehabilitation centres, etc.). Arguably, mixed approaches to renewal (e.g. Rotterdam, Barcelona) have been more successful in broad social terms than mainly market-led strategies (e.g. London Docklands) but one of the fundamental problems facing such cities remains the often poor quality of social, health, education and security provisions, and the financial problems many cities face in making improvements in basic services. Although some inner areas have been upgraded through inflows of higher income residents ('gentrification') the general pattern is still one of continuing suburbanisation of jobs in back offices, retailing, distribution and manufacturing. Since cities depend heavily on property, residential or employment taxes, they face the long standing local taxation problem (Eversley, 1972; CEC 1991b: 147) of rising costs and static (or falling) incomes and may require increasing levels of subsidy from central government to overcome problems of spending and provision in a range of services from hospitals and schools, to policing and social services and transport infrastructure.

Less favoured rural areas

As a whole, rural areas in all parts of the community have had to face considerable changes over the past few decades. Employment in farming has continued to decline due to the application of modern technology and more capital intensive methods, and crop yields have risen fast, greatly increasing the levels of output achievable on existing farmland, producing considerable market surpluses in many products by the early 1980s. Subsequently, market forces have been given greater prominence in reforms of the Common Agricultural Policy (CAP) system of intervention and price support, with an inevitable depressing effect on market prices. In addition, compensation has been provided for farmers to take part of their land out of production. The policy of set-aside was expected to result in a reduction of farmland of around 15 per cent. Future improvements in farm incomes are likely to arise from a hardening of prices as markets move nearer balance, continued improvements in farm structures and sizes, and improved job opportunities in rural areas, enabling more small-scale farmers to supplement their incomes by going 'part time'. Nevertheless, the combined impact of productivity improvements and a decline in the number of smaller farms, is largely responsible for employment reductions in this sector running at 3–4 per cent per annum over recent years, and almost a 50 per cent reduction over the period 1973–93 (Jones, 1996).

However, some rural areas have faced much greater problems than others and this is especially the case with mountainous and hill-farming zones and other less favoured areas covering most of Spain, Portugal, Italy and Greece, Scotland, Wales and the west of Ireland, parts of northern and south-west England, south-west France, and large parts of Germany (Blacksell and Williams, 1994). With the addition of Austria, the European Union added a large alpine region and large parts of Finland and Sweden benefited from 'Nordic support' involving long-term compensatory national aids for farms situated north of 62 degrees north and adjacent areas which suffer from a short growing season and the isolation of border regions. In addition, a new Objective 6 for the Structural Fund was brought in for regions with a population density of under 8 inhabitants per km^2. However, in the new Member States only Austria's Burgenland qualified as an Objective 1 (i.e. lagging) region.

There is a considerable overlap in spatial coverage between the problem rural regions, and the peripheral regions of the EU. This generally implies that such areas can expect to be more distant from important financial and commercial centres and for manufacturing to represent a much lower share of total employment. Although these factors may make it more difficult to attract new investment, rural areas with attractive natural environments such as coastal, wooded or mountain areas have frequently achieved substantial employment gains (e.g. French Atlantic Coast, Black Forest), and areas such as the French Southern Alps and south-west England have seen population inflows through retirement migration (CEC, 1994d: 122). Some areas will be able to improve their attraction for industry by improving their transport and telecommunications links, and others will need to concentrate on environmental improvements and tourism and leisure facilities. Areas experiencing residential decentralisation or retirement in-migration may question the overall benefit to existing communities, and greater direction from longer term and strategic planning may be required to achieve a better balance. Support for rural towns (key settlements in UK planning terminology or *villes rurales* in France) is once again becoming popular, both to ensure and improve access to a wide range of local services and to act as a local focus for economic development (CEC, 1994c; *Le Monde*, 1997).

It seems inevitable, however, that some of the more remote rural regions, with weak agricultural sectors, will continue to see the decline of agricultural and rural populations if diversification proves difficult, although such areas may conceivably benefit the most from other CAP reforms to promote less intensive farming methods and encourage conservation.

Border regions

Prior to the accession of Sweden, Austria and Finland, the European Union had some 10,000 km of borders, 40 per cent of which were external and 60 per cent internal. Around 10 per cent of the EU's population live in these areas (CEC, 1994c: 125). Many now internal border regions were at a considerable locational disadvantage prior to joining the EC, being inaccessible from the main national markets and with relatively restricted trade flows with adjoining countries (e.g. French Pyrénées or 'old' West German border regions). They frequently also suffered from poor cross-border transport infrastructure due to the historical separation of national transport networks. These areas are therefore likely to benefit considerably from the development of the trans-European networks (TENS), which are attempting to close up 'missing links' in cross-border and international transport infrastructures (e.g. Channel Tunnel Rail Link). In some cases this may also imply major adjustment to increased competition and changing locational factors in areas that may have previously experienced little trans-border competition.

Estimates suggest that cross-border trade shortly after 1992 was growing by an average of 0.4 per cent a year faster than the EU economy as a whole (Verchére, 1994), so regions and cities located near strategic cross-roads should benefit considerably if they are able to develop as international production, distribution, service or financial centres.

Trans-border co-operation can help support a common approach and reduce duplication of resources in a wide range of areas of common interest such as trade and tourism, transport and communications, the environment and rural development and this has been encouraged through the Community Initiatives such as INTERREG. Projects in border regions may also benefit from other structural funds, even where they are not designated areas as such (i.e. Objectives 1, 2 and 5b).

Peripheral regions

Regions situated at the periphery of the EU are often at a considerable distance from the main markets and populations in the SEM and may consequently rely largely on national markets or even more localised regional centres of demand and supply (e.g. Madrid, Lisboa, Athinai). Firms thus find it difficult to develop to a size where they can achieve economies of scale, and are simply unable to compete with large multi-national companies located in more central regions (Armstrong and Taylor, 1993). In one study of Northern Ireland manufacturing firms, three-quarters stated that they were disadvantaged because of the area's peripheral location and pointed to the main problems as being: higher transport and input costs; communications with suppliers; marketing and selling products; keeping abreast of changes in the industry; peripherality inhibiting the emergence of industrial clusters and thereby reducing scale and agglomeration economies; difficulties in finding qualified workers in specialised sectors and access to technical expertise (Sheehan, 1993).

In addition, and probably contributing to some of the above-mentioned problems, levels of infrastructure provision have been found to be generally deficient in peripheral southern and western regions, whilst provision was found to be above average in the centrally situated and more developed areas (CEC, 1987). More recent studies have pointed to significant gaps in transport infrastructure, telecommunications, energy and education, and estimates suggest that over ECU 18.5 billion would have to be spent in lagging Objective 1 regions (mostly peripheral) up to 2005 just to fulfil environmental commitments concerning the disposal of urban and industrial water waste and urban and industrial solid waste (CEC, 1993a).

The European Commission originally defined peripheral regions by estimating their 'economic potential' (EP), that is with reference to eco-

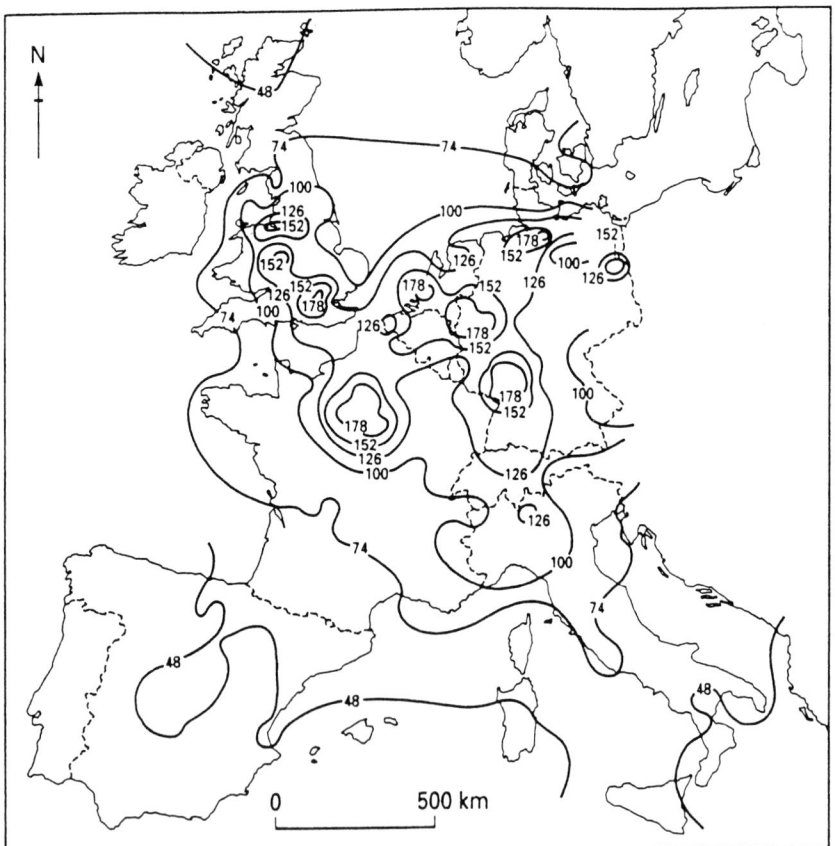

Figure 2.2 Economic potential, EU, 1983
Source: Drake (1994)

nomic activity (regional GDP) in each of the regions and to the distances separating them. In calculating EP for any region a gravity model is used so that regional GDP in all other regions is first divided by the distance separating them. Thus central regions tend to have high EPs as they are generally close to many other high income regions, whereas for opposite reasons, peripheral regions tend to have low EPs (Figure 2.2). Peripheral regions were defined as having an EP below 60 per cent of the EC average – this included four Member States in their entirety (Ireland, Greece, Spain and Portugal) together with Northern Ireland, Scotland, the North and extreme South-West of the UK, parts of Denmark, Sud-Ovest (Italy), and Sub-Ouest and Corse (France). These regions covered 55 per cent of the EC territory but accounted for only a third of its population, 29 per cent of employment and less than a quarter of its production (GDP). They accounted for over 55 per cent of the agricultural employment in the EC, almost 40 per cent of the unemployed and had regional incomes (GDP per capita of population) of less than 75 per cent of the EC average (CEC, 1987). By the early 1990s the position had changed little except that although employment had risen slightly faster than the EU average since 1986, unemployment had increased both in absolute terms (to 16.7 per cent) and relative to non-assisted areas (8 per cent) (CEC, 1994d: 121).

Following the Structural Fund reforms, Objective 1 (or lagging) regions are characterised by per capita income levels below 75 per cent of the EC average. In a policy context there is a clear overlap between peripherality and the officially designated lagging regions, although peripheral areas of northern Spain and south-west France with per capita incomes above 75 per cent of the average were not designated, and neither were parts of north-east Italy. A few areas suffering from industrial decline were included – Merseyside and Hainaut in Belgium – although the latter is hardly periph-eral. Parts of the Austrian Alps as mentioned earlier now also qualify. Elsewhere in the UK only the Highlands and Islands and Northern Ireland were classified with Objective 1 status. It is probable that various forms of inter-regional transfers through, for example, social welfare and infrastruc-ture spending have helped to raise slightly regional incomes in the periph-eral areas of France, Italy and the UK, and in the new Member States Sweden has consistently made efforts to integrate and compensate indivi-duals and regions in the far northern part of the country. The addition of Finland and Sweden will have changed the EP map only slightly due to the small population size of these countries and the considerable distances involved – however it seems likely that, at the very least, most northern parts of these countries would be regarded as peripheral. The inclusion of the former East Germany has added a region which, with an initial per capita income of less than 35 per cent of the EU average, is nevertheless located close to many of the high income central regions of Europe, and therefore its level of peripherality, if any, is unclear at present. So far as the future addition to the EU of Estonia, Hungary, the Czech Republic,

Slovenia and Poland is concerned, it seems likely that all that can be said is that whilst some areas will be closely linked to the central core regions of Europe, particularly as the transport infrastructure improves, Estonia and parts of Poland, Hungary and possibly Slovenia will almost certainly emerge as being peripheral. It also seems probable that the expansion eastwards and northwards of the EU will lead to some relative deterioration in the EP of existing peripheral regions in the south, west and north-west (although for the UK this may be offset by improvements in access to European markets through the Channel Tunnel and Rail Link).

Finally, the overall demographic and labour market situation influencing the peripheral regions needs to be considered. The Commission has drawn attention to the fact that to some extent the problem of high rates of unemployment in the less developed regions is related to demographic trends. Higher birth rates continue to result in faster growth in the labour force than elsewhere in the Community. Stronger employment growth is therefore needed in regions lagging behind to offset the relatively faster growth of the labour force before unemployment disparities with the rest of the Community can begin to be reduced (CEC, 1991a: 12).

More recent estimates for the 1990s echo the same view: the projected 4 million increase in the working age population (15–64) would mainly be concentrated in the south of the EU and in Ireland (CEC, 1994d: 25). By contrast, the labour force in parts of Germany and some of the more urbanised and central parts of the EU could actually decline. Of the seventy-nine regions with above average rates of unemployment, around two-thirds can expect to see a faster than average growth in labour supply over the latter part of the 1990s (CEC, 1994d: 29). It is unlikely that out-migration of workers from peripheral areas would have much effect in greatly reducing the impact of such labour market imbalances. In the first place, labour mobility is much lower than in the past. During the 1960s net out-migration from 'southern and western' peripheral regions averaged over 15 per thousand (‰) per annum, but by the 1980s this rate had fallen to 0.7‰ (Champion *et al.*, 1996). Within Europe there are significant cultural, language and financial impediments to labour mobility. Regional house price differences, for example, are often considerable and the volume of cheap rented accommodation in many European cities has declined greatly over recent years (Bull, 1996). The growth of dual-income households, and indeed owner-occupation has significantly raised cost barriers facing potential migrants (Blotevogel and Fielding, 1996). To the extent that migration is more of a 'pull' process than a 'push' process, it probably has more effect on attracting workers to areas of job expansion than on encouraging workers to move in response to regional differences in unemployment per se. During the 1960s employment growth in central industrial regions was high, as was migration, but in the 1990s unemployment in central regions is itself high, (albeit lower than at the periphery), hence the pull effect on migration is lower. Within the southern periphery, however, relatively localised rural-urban migration has fuelled the continued growth of many cities such as Madrid, Sevilla, Napoli, Palermo and Toulouse (CEC, 1994d: 97).

In order to prevent growing regional disparities in unemployment there is a considerable need to encourage new investment and relocation of firms to Europe's problem areas and, in particular, to lagging peripheral areas. However, although labour costs in the regions of Spain or Portugal are probably less than half those in Germany, they are nevertheless still considerably higher than in developing countries and probably four times higher than in Hungary or Poland (*Financial Times*, 7 June 1993). In addition, the advantage of lower wages may be partially or wholly offset by lower labour productivity, itself due to poor infrastructure, lower skills and training of the workforce and a lack of scale and agglomeration economies. Other cost factors may be related to distance from main European markets. Unit labour costs, which take account of both productivity and labour costs, may consequently be above average in peripheral areas even though wages there may be lower. Areas which gain a productivity advantage tend to maintain it, since although there may be some increase in wages, as a whole labour costs differ less between regions than productivity, thereby reducing unit labour costs in high productivity growth areas. The opposite can occur in lower productivity regions; for example, studies found that regions in southern Italy, Greece and certain areas with structural problems in the northern parts of the European Community, all had higher than average unit labour costs (CEC, 1987). In conclusion, policies towards peripheral areas must not only improve infrastructure and accessibility in a whole range of areas, but they also need to widen the industrial base and improve productivity through investment in expanding sectors and indigenous firms. The rural economy is often an important part of many peripheral regions and represents another diverse set of problems.

Growth and adjustment in the regional economy – convergence or divergence?

The history of integration shows us that the process of regional adjustment can not only be long and difficult but may also be cumulative in the sense that the relative economic position of negatively affected regions may actually continue to worsen over a considerable time-scale. One of the earliest examples in Europe followed the unification of Italy after 1861, when substantially lower (Piemontese) external trade tariffs were imposed on what was then a very protected but nevertheless industrialised southern economy. Until the first real efforts to promote industrial development in the late 1950s the economic history of the area (The Mezzogiorno) was one of uninterrupted decline (Allen and Maclennan, 1970) and policy encouraged out-migration. The problems of this region are today still considerable, as illustrated by the presence of some of the highest investment incentives in the EU (Yuill *et al.*, 1996).

Models which support the idea of cumulative growth (and decline) were initially developed by Mrydal (1957) and Hirschman (1958). They were based around the assumption that economies of scale and agglomeration in 'core' central regions would provide substantial ongoing and even increasing benefits to firms locating there. Resources – including capital and labour – would be drawn to these rapidly expanding central regions, increasing these economies even further. Firms would continue to be drawn towards such central regions due to their proximity to major market centres, an abundance of supporting service activities and proximity to major national and international transport and distribution facilities and networks. Peripheral regions would see some benefit from increased demand for raw materials and food production and areas surrounding the 'core' would experience some 'overspill' investment, but, in general, peripheral areas would fail to develop economies of scale and agglomeration in the same way, and would lose out due to the selective migration of skilled workers and the 'syphoning' off of regional savings through capital mobility in the banking system. Under such 'cumulative causation' models the richer core regions could well continue to become richer whilst peripheral regions might in fact become poorer as existing industries failed to compete with larger, more economical 'core' producers. Diseconomies might eventually arise in 'core' areas through excessive congestion and pollution, but in spite of high social and economic costs, firms might resist moving to peripheral locations. The theory was added to by Kaldor (1971) who suggested how wage adjustment lags in labour markets might further encourage this process and Dixon and Thirlwall (1979) who developed an export-base model of regional growth.

Neo-classical approaches to this problem rely on inter-regional movements of capital investment and labour to help achieve regional balance. In weaker regional economies (with lower productivity) or where labour market growth is excessive, workers are expected to accept lower wages. Firms are expected to react to this relative lowering of costs by relocating production to these areas and local firms are expected to become more competitive and thereby increase investment and productive capacity. Workers will react by seeking jobs in higher paid regions and out-migration will increase.

In Figure 2.3 there is initially both a fall in demand for labour and a rise in supply due to the factors outlined above (to D_2 and S_2). The equilibrium regional wage falls from W to W2. Firms raise investment in the region (shifting demand back up towards D_1) and workers migrate elsewhere (lowering regional supply of labour back up towards S_1). Equilibrium is expected to move up towards E_1 and regional wage rate differences will therefore narrow (under conditions of perfect factor mobility, wage rates would actually equalise as between regions).

The general view, however, is that labour mobility within Europe is

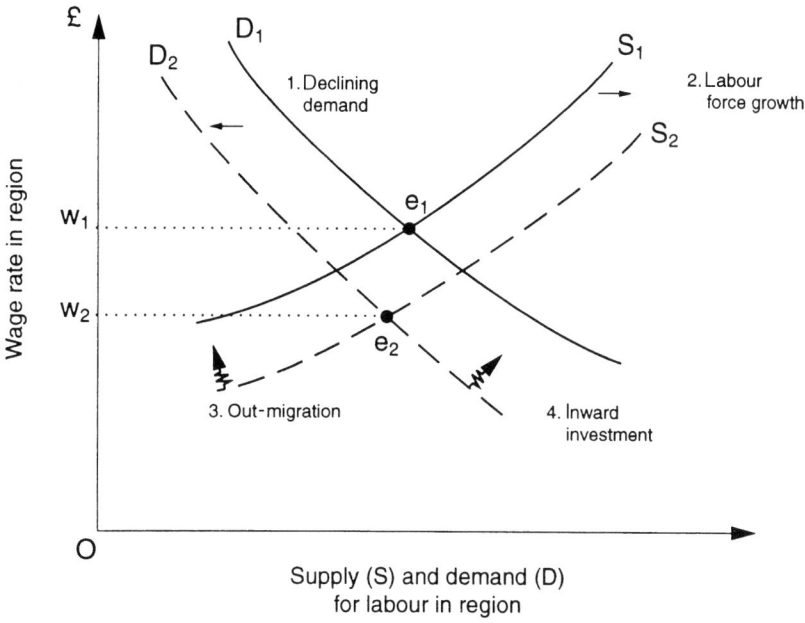

Figure 2.3 Effects of shifts in demand and supply on regional labour markets

rather low and perhaps only a third of the level in the US (Maclennan and Stephens, 1997) – although direct comparisons are difficult, particularly since migration in the US has largely been determined by international (predominantly South American) rather than internal movements.

It is also doubtful whether large-scale migration would ever be beneficial for depressed regions, especially if it took the form of an exodus of more skilled and better educated workers (CEC, 1994d: 150) and excessive migration could raise demand for public utilities and infrastructure in already congested regions whilst leading to cuts in provision in depressed regions. Geographical distance and the depressed labour market in many of the central regions of Europe are also reasons to expect that, with the possible exception of new Eastern European members, migration within the EU will remain low.

Capital mobility could in theory offset the effects of low labour mobility from depressed regions and there is some evidence of increased geographical mobility. Where 70 per cent of employment was probably tied to its existing location in the 1950s and 1960s, this has now fallen to around 50 per cent (CEC, 1991b:15). Certainly there is some evidence that firms may raise profits by relocating from congested and high cost urban locations. Tyler and Moore (1988) for example, showed how profitability in the majority of manufacturing industries could be increased by more than 20 per cent by a move of at least 100 miles from inner London. Most of these gains were estimated to come from lower wages and salaries and the lower cost of industrial services, but this takes no account of the availability and quality of these factors. The study did suggest that many less urbanised areas in relatively close proximity to London and Birmingham might offer similar cost advantages to the more peripheral regions, and it seems reasonable to assume that access to business services and skilled labour here would often be much better than at peripheral locations, given the generally high spatial concentration of such services (O'Farrell *et al.*, 1996). In conclusion, if firms do decentralise they may not need to go far to receive a satisfactory reduction in costs. Long distance moves to peripheral areas, unless carefully considered, could result in unexpected relocation costs associated with changes in suppliers and distance from main markets, as well as shortages of skilled labour and quality of business and financial services.

Capital mobility has also taken place through a considerable increase in foreign direct investment (FDI) between EU Member States. Between 1986 and 1991 there was around 150 billion ECU of FDI between Member States and a further 117 billion into the EU from elsewhere. Apart from Greece, FDI in Spain, Portugal and Ireland has been even greater in value than EU regional aid (CEC, 1994d). From 1992–94 EU Member States invested over 8 billion ECU in the countries of Central and Eastern Europe (CEE) – a similar amount to EU FDI in South America, the newly industrialised countries (NICs) of Asia, together with China (CEC, 1997a). Given its low starting base, FDI in CEE countries seems set to rise considerably in the future, but will need to do so in order to help offset likely and probably significant adjustment costs in these countries.

However, much FDI, especially that originating from outside the EU, does not in fact reach the peripheral regions. Where it does, investment tends to cluster in a few areas (e.g. Central Scotland, mid-west Ireland, Puglia (Italy), Spanish Mediterranean Coast, Thessoloniki in Greece) with fairly well developed socio-economic infrastructure (CEC, 1994d: 83). A recent survey of FDI in France in 1996 showed that of 22,800 jobs attributable to FDI, over 56 per cent went to regions along the eastern Belgian, German and Italian frontiers and only 12 per cent to regions in the west and south-west from Brittany to the Mediterranean (Menanteau, 1997). Whereas 60 per cent of jobs in the above example came from other EU countries, FDI in the UK (CEC, 1994d) mainly comes from non-EU countries (USA, Japan, NICs, etc.) establishing a 'bridgehead' in Europe.

Firms undertaking such investment are often themselves large multi-national corporations operating on a global scale. They are more likely to be involved in high-tech sectors (but not necessarily high-tech production). Such sectors are characterised by high levels of intra-industry trade, where economies of scale are likely to be gained in the production and exchange of highly differentiated products. On balance, such newer types of specialisation may actually contribute to a widening of regional disparities, particularly if internal and external economies can be maximised close to core high-density regions of the Single Market. The large size of firms involved may also enable them to establish dominant positions in national markets and thus circumvent normal competitive pressures (Armstrong and Taylor, 1993). With relative ease of access to expanding (peripheral) markets, Eaton and Lipsey (1979) showed that it would theoretically pay existing market leaders to establish new plants before the time it would first pay new (local) firms to enter.

Indeed, some authors have questioned whether trade liberalisation under conditions of economies of scale would necessarily result in optimal specialisation, as trade may go to the advantage of the original dominant producer who may not be the most efficient (Grubel, 1967).

Trade advantages by producers in peripheral regions may nevertheless be gained in sectors where they have a comparative advantage (CEC, 1990a) derived from factor endowments (e.g. relatively cheap labour or raw materials) and where unexploited economies of scale exist (e.g. textiles). In the case of intra-industry trade, advantages may arise in differentiated products where regions are able to develop a regional specialism. Finally, in the case of multi-national branch-plant location, according to Fielding (1989) this has increasingly come to be characterised by the New Spatial Division of Labour (NSDL) which involves the growing geographical separation of command and head office functions, including R&D, from routine production and assembly processes, for which more dispersed and lower cost locations are usually sought. Areas such as Wales and the north of England have been particularly successful in attracting this type of FDI (Hill and Munday, 1992). Stone and Peck (1996) estimate that by 1993 around a third of manufacturing employment in Wales was in foreign-owned plants. In Northern Ireland the figure was more than one in four and in Scotland and the North more than one in five, compared to a national figure of around one in six. These authors also point out that at least as much job gain in the foreign-owned sector was due to acquisition of existing UK-owned plants as to openings of new factories, and that if the employment effects of the former are excluded, Northern Ireland and Scotland would actually have seen a net decline in foreign manufacturing employment over the period 1978–93. Phelps (1997: 156) points to the fact that little of the multi-national manufacturing activity in Wales is embedded to any significant degree, and that policies and grants aimed at inward investors have had limited impact on the nature and embeddedness of their investments. Nevertheless, there is some evidence that multi-nationals may increasingly be looking at qualitative factors such as environmental and cultural aspects and good access to a skilled and well qualified workforce. In this context, Phelps (1997) points to the problems which areas like South Wales are likely to experience given that important reservations were recently expressed by a major inward investor concerning the growing skills shortage in Wales. One of the problems here is the inadequacy of national educational and training systems and the fact that it is probably almost impossible for local development and training bodies to offset such basic supply-side inadequacies at a regional level (Phelps, 1997: 158).

It is also the case that many peripheral regions are sited well beyond the possible spread of developing *technopôles* in continental Europe, covering southern Germany, south-east France, north-west Italy and north-east Spain (Harrop, 1996). These rapidly expanding areas, and particularly those in France, are good examples of how long-term strategic planning backed up by transport and infra-structure improvements and the development of policies towards research and development (R&D) can achieve a high degree of success. Holland (cited in Harrop, 1996) suggests that since the likelihood of *technopôles* arising spontaneously in problem regions is very remote, the EU Structural Funds should help encourage the establishment of new R&D centres in less developed regions.

To the extent that cumulative causation models are more spatially explicit than neo-classical models, they are of more help in understanding the long-run nature of regional problems and the persistence of regional disparities. Clearly, although regions can overcome many of the problems caused by peripherality or declining industries (e.g. Ireland is now one of the fastest growing EU Member States), there is nevertheless a real possibility that some regions or sub-regions may continue to diverge. Such occurrences may easily be masked by overall statistical measures of convergence and divergence; it is perfectly possible, for example, to obtain a measure of overall convergence whilst some proportion of problem regions become poorer and others move closer to the norm. Furthermore, the way regions are defined and the growing number of regions in an expanding Europe will also make it more difficult to assess long-term regional trends.

Economic and Monetary Union (EMU)

As discussed earlier, the Delors Report set out the regional stages towards EMU and the precise timetable and conditions and convergence criteria were set out in the Treaty on European Union in Maastricht (1991). Whether more than half a dozen countries will meet the convergence criteria by 1999 is, at the time of writing unclear. However, countries joining at a later date will clearly be affected by these events as (unless they have an opt-out) they are still required to maintain macro-economic convergence and stability targets before they eventually join.

In general, the Commission expects a single currency to improve the operation of the SEM which is clearly undermined by the existence of different currencies and by substantial currency movements. These not only produce transaction costs (about 0.4 per cent of GDP) in changing currency, but may also reduce dynamic gains from the SEM by impeding investment and location decisions of firms unsure of future currency risks. Overall, the existence of transaction costs acts as a brake on trade and

Table 2.4 Aspects of economic and monetary union

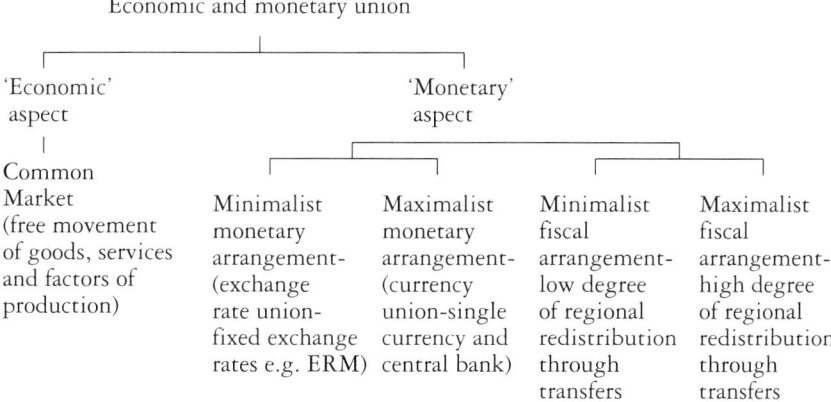

Economic and monetary union

'Economic' aspect	'Monetary' aspect			
Common Market (free movement of goods, services and factors of production)	Minimalist monetary arrangement- (exchange rate union- fixed exchange rates e.g. ERM)	Maximalist monetary arrangement- (currency union-single currency and central bank)	Minimalist fiscal arrangement- low degree of regional redistribution through transfers	Maximalist fiscal arrangement- high degree of regional redistribution through transfers

Source: Adapted from Swann, 1996

specialisation, as well as competition and consumer benefits from greater choice and price reductions.

The 'monetary' aspect of EMU can be broken down into varying monetary and fiscal arrangements (Table 2.4) with greater and lesser degrees of integration. Opponents of EMU tend to argue that 'economic' rather than 'monetary' union is where most of the gains can be achieved.

Starting from a minimalist position, this would involve a commitment to fixed exchange rates and full convertibility (without restrictions) of currencies. Minimalist fiscal arrangements would involve only limited redistributive power of the Union budget in terms of revenue and spending and this would probably also imply a relatively small overall budget. While such an approach might seem desirable from the point of view of maximising national sovereignty, there are a number of shortcomings. First, it may prove difficult to develop policies and spending to offset any existing or resulting regional imbalance or to offset the regional effects of other policies (e.g. agriculture) or even to develop policies of common interest (e.g. research, energy, transport). Second, there would still be a considerable degree of monetary and fiscal discipline required simply to maintain fixed exchange rates and it could then be argued that there would be little additional cost – but considerable trade benefit – in irrevocably fixing exchange rates and then replacing them by a common currency.

This would occur under the 'maximalist' monetary position where a single currency would be accompanied by a central bank responsible for controlling the supply of the common currency, influencing lending and interest rates and management of the external value of the currency (it would control gold and foreign exchange reserves). In theory the central bank could either be independent or it could take instructions from political authorities and its aims might involve price stability or employment

(Swann, 1996), although it is difficult to see how it could always achieve both. Although national governments would still set levels of public spending in most areas as they do now, they would almost certainly find some limitation on the size of national budgetary deficits (in the interests of monetary stability in the Union). The determination of overall monetary and fiscal policy within the Union would therefore become a complex and highly politically-charged issue.

Maximalist fiscal arrangements would also achieve a high degree of automatic fiscal stabilisation, reducing the tax-take from, and increasing public expenditure transfers to problem regions if economic activity declines. Much of this redistribution occurs within national economies through tax and social security systems, although Federal systems such as in Belgium and Germany often have inter-regional fiscal equalisation mechanisms. For example, in Germany there is a redistribution of VAT receipts to the weaker *Länder* (CEC, 1994e: 163). The degree of regional stabilisation or 'fiscal offset' afforded by public finance channels (i.e. the degree to which it is able to offset changes in local economic activity) is probably between 17–28 per cent in the US, compared to around 34 per cent for the UK and between 33 per cent and 42 per cent for Germany (CEC, 1993a). The redistributive effect of the current EU budget is by these standards very small but it has nevertheless become quite important for certain local and regional authorities. The European Commission estimated that annual transfers throughout the Cohesion and Structural Funds had added around 2.3 per cent to the GDP of Greece, Portugal, Ireland and Spain and represented around 6 per cent of investment in these countries (CEC, 1994d: Table 20). Nevertheless, the degree of 'fiscal offset' is undoubtedly low and Minford (1997) puts forward a maximum figure for any region of less than 1 per cent.

For the problem regions of Europe, EMU is likely to produce costs as well as benefits. The need for tightening of monetary and fiscal policy to achieve strict convergence criteria has undoubtedly produced short-term adjustment costs, which have been all the more difficult because of the general recessionary climate of the early mid-1990s. The Commission's annual economic report puts annual growth at only 1.5 per cent per annum from 1991–96 and records a loss of some 4.5 million jobs. In addition, countries and regions in weaker countries would no longer be able to devalue their currencies to cushion against the effects of external competition or a sudden rise in costs. Furthermore, Minford (1997) suggests it may actually prove more difficult to stabilise fluctuations in the economies of the EU under a single currency. Against this, producers and consumers will benefit from improved access to the SEM, reduced transaction costs and, where weak currencies have had to bear exchange rate risk premiums, the benefit of lower interest rates.

In today's global economy characterised by a high degree of deregulation of international finance and foreign exchange, the scope for governments in

EU countries to run an independent monetary policy is severely proscribed. Around 90 per cent of international exchange transactions are now speculative (as opposed to being trade or investment based) compared to only 10 per cent in 1971, and the volumes involved are now such that national central bank reserves are unable to cope with large speculative outflows (Eatwell, 1995). This effect could be seen working in reverse in the UK in 1997 when the newly independent Bank of England decided to raise base interest rates on several occasions to curb inflation, and foreign exchange markets quickly pushed the pound sterling up to eight year highs against other EU currencies. The impact of such large currency fluctuations on trade and investment within Europe could be large enough to offset many of the benefits of the SEM and the effects on particular regions specialising in exports could be particularly severe. Given that most of Britain's trade now takes place with other EU countries (in 1995 the UK exported more to the Netherlands than to China, South Korea, Hong Kong, Indonesia and the other Asian 'tigers' combined), it can be argued that stability within the EU is now more important and should be given relatively greater emphasis than other external and internal considerations in domestic policy. This stability would be achievable if Britain were to join the single currency. Although some compromises would be inevitable regarding, say, interest rates or public spending within the EMU, it can be argued that nationally any costs would be relatively small compared to the benefits of freedom from major exchange rate movements. In addition, the 'casino' model of world capital flows which has given financial speculators enormous influence (Lovering, 1997) would be much reduced by the introduction of a single currency, which would effectively produce a world currency (the Euro) to rival both the dollar and the yen. Last, London's position as Europe's largest financial centre could be in some doubt if Britain did not join the single currency.

Conclusions and future trends

So far European integration has produced considerable overall benefits – in 1996 the Commission estimated that the SEM had been responsible for creating up to 900,000 extra jobs and lowering inflation by 1–1.5 per cent (EC, WE/38/96). Yet overall the recent poor performance of the EU economy means that regional adjustment processes following the SEM and EMU are likely to be all the more difficult. Given the lack of political consensus within the EU it seems unlikely that a system of fiscal transfers will emerge to act as a safety net for Europe's problem regions, and such regions are therefore largely dependent on national redistributive systems and levels of fiscal offset. This could represent a particular problem within the regions of the weaker Member States. Although the Cohesion Fund was established by the Maastricht Treaty to offset the costs of convergence towards EMU among these states (amounting to ECU 15.5 billion over the period

1994–99) the sum involved was undoubtedly insufficient to cover the expenditure required for this purpose.

Even in terms of regional policy it is probably still somewhat misleading to talk of a European regional policy, and there is still a highly decentralised process of decision-making and implementation, whilst the power of the Commission is still relatively limited (Tsoukalis, 1997). It therefore seems unlikely that the Commission will be able to achieve a completely level playing field for regions attempting to attract internationally mobile investment, and wasteful competitive outbidding in regional aid seems likely to continue until (or if) some harmonisation of financial incentives, corporation tax rates, etc, is achieved. Large national variations in these areas as exist are hardly compatible with the operation of a single market.

It seems difficult to ignore the undoubted benefits that a location near the central 'core' regions of Europe will bring, and even in sectors such as financial services there is evidence to suggest that whilst a more explicit European hierarchy may develop (with gainers and losers) the benefits will accrue disproportionately 'to existing financial centres which tend to be in the more prosperous and rapidly growing regions of the European Community' (Hardy *et al.*, 1996). Overall there is no reason to doubt the conclusion of the Committee for the Study of Economic and Monetary Union (1989),

> Historical experience suggests, however, that in the absence of countervailing policies, the overall impact on peripheral regions could be negative. Transport costs and economies of scale would tend to favour a shift in economic activity away from less developed regions, especially if they were at the periphery of the Community, to the highly developed areas at its centre.

In addition to traditional locational factors, the role of accessibility to rapidly developing European axes of economic development is likely to be of growing importance for a number of reasons. First, most firms are likely to experience a growing spatial market as sales and supplier networks are expanded and accessibility therefore becomes more important. Second, large firms in the process of rationalising production in the SEM are more likely to concentrate production on fewer plants but with higher, possibly Europe-wide trade flows. Third, as we have seen, cross-border trade is increasing rapidly and new international axes are thereby developing. These patterns are becoming clear from models such as that developed by IAURIF (1991) (see Figure 2.4) together with the now well-known 'blue banana' model developed slightly earlier by the French planning agency DATAR (see Figure 2.5). The latter was itself very similar to a model developed much earlier around a 'Manchester-Milan' growth axis. More recent attempts to define European 'super regions' appear to have little economic coherence and have met with some scepticism (Gripaios, 1995).

Figure 2.4 International axes of development in the EU
Source: IAURIF (1991)

Overall, future patterns of spatial accessibility will be greatly influenced by the development of trans-European transport and telecommunications networks (TENS) supported by the Structural and Cohesion Funds as well as the European Investment Bank (EIB) and the European Bank for Reconstruction and Development (EBRD). In the future, the Structural and Cohesion Funds are unlikely to see any major growth – a figure of 275 billion ECU has been put forward by the Commission for 2000–06 compared to 200 billion ECU for 1994–99 (all at constant 1997 prices), with a figure of 45 billion ECU set aside to facilitate enlargement. Overall, support would continue to represent around 0.5 per cent of EU GDP, and the maximum level of transfers would not be allowed to exceed 4 per cent of the GDP of any current or future Member State (CEC, WE/28/97). Operation of the Fund is likely to be simplified and decentralised, largely on the basis of the principle of subsidiarity. It seems

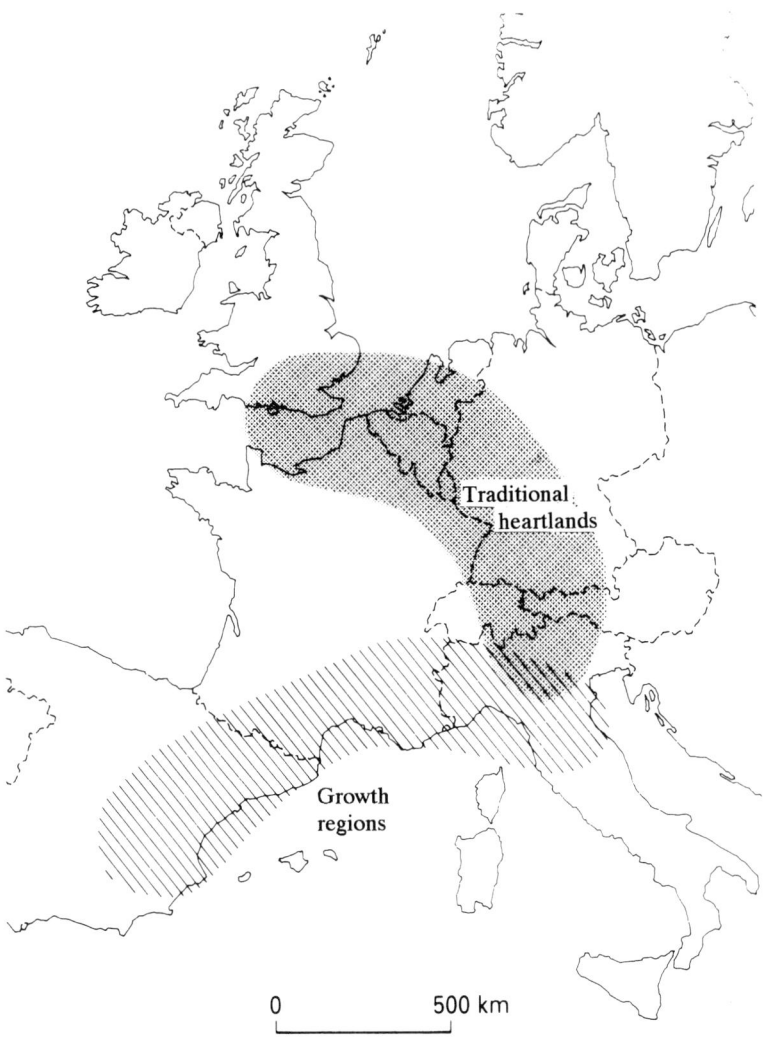

Figure 2.5 The traditional heartlands or *blue banana* model of economic development
in the EU

probable that the Commission will be less involved in the detailed implementation of Structural Funds. Funds may be linked more directly with
national policies and programmes and the focus will be on agreeing overall
policy frameworks, leaving implementation of the programmes to the
relevant parties within each country (CEC, WE/13/97). Whether this

provides an expanded role for local and regional authorities remains to be seen, and may depend crucially on institutional factors within each country. However, there is clearly no discernible trend towards a 'European' regional policy; perhaps an inevitable outcome of the growing diversity in an expanding EU. The establishment of a single programme for each region would, if it were realised, greatly simplify the perhaps confusing array of aids and initiatives from Brussels with which authorities must currently deal.

3 Introduction to government and planning systems

There is a widely held view that regional planning in Europe has evolved within five very distinct legal and administrative frameworks: British, Napoleonic, Germanic, Scandinavian and East European (Newman and Thornley, 1996). The British legal system derives from a combination of English Common Law (which operates in most of the United Kingdom and in Ireland, and is based on the empiricism of case by case decisions) and a hybrid mixture of local customary law, Roman Law and English Common Law in Scotland. Britain is also alone in Europe in having no written constitution, and hence no special protection in law has been granted to local government, while in Ireland (because of the legacy of British dominance) local authorities are likewise unprotected in law. Whereas in much of Continental Europe – particularly within the Federal states – there is an assumption that local and regional authorities 'have a general power over the affairs of their communities' (Newman and Thornley, 1996: 30), in the United Kingdom and Ireland – notwithstanding the need to provide public services on a local scale as agents of central government – responsibility so to do is subject to the concept of *ultra-vires* if the local authority exceeds the powers conferred upon it by the centre. Clearly, the administrative system of the United Kingdom and Ireland is a dual system (see Leemans, 1970; Bennett, 1993; Newman and Thornley, 1996). Local authority activity and finance are constrained by central government, while central government exercises a supervisory role. Duality is also manifested by little movement of politicians and professionals between local and central government (or *vice versa*) and there are very few instances of politicians holding office at more than one level of government – in contrast to practice in many other European countries. With the gradual devolution of government in the UK to Scotland, Wales and Northern Ireland, and possibly to the English regions, the dominance of Westminster and Whitehall might begin to break down as power and responsibility move closer to the electorate.

In contrast to the empiricism of the British legal style, the French Civic Code of 1804 – based on abstract principles – provided the model for all codes of private law throughout those areas of Europe incorporated into the

Napoleonic Empire. After independence, post-1815, the basic elements of the French (or Napoleonic) Code were retained across much of Europe – notably in Belgium, the Netherlands, Luxembourg, Spain, Portugal and Italy, as well as remaining largely intact in France to the present day. Within Napoleonic Europe, the structure of local government, and the degree of power attributed to it, contrasts markedly with that of the British administrative family. Whereas in the United Kingdom and Ireland fairly large local authorities have emerged to facilitate the provision of efficiently managed services, administrative systems in much of French-influenced Europe place much importance on the local commune at the lowest level of government – in part a result of the centuries-old administrative struc- ture of the Catholic Church, and (in France at least) as an attribute of the revolutionary democratisation process. However, to ensure that the central state remains paramount, while at the same time local government performs its agency role in the provision of services, a system of administration has emerged whereby career civil servants from central government (préfects) are deployed locally to facilitate a fusion between strong central control and local representation (Leemans, 1970; Bennett, 1993). In Spain and Italy, the development of a degree of regional autonomy in recent years has broken down the 'Napoleonic' relationship between the upper and lower tiers of government, while in Belgium Federal structures of government have emerged, all but superseding the French Code.

Unlike the unified British and Napoleonic legal systems, the Germanic Code – rather than being imposed by a central power – evolved within a patchwork of German states during the fifteenth, sixteenth and seventeenth centuries, and in accordance with the legal concepts and institutions of Roman Law. Although subsequently sharing with the Napoleonic family an adherence to codification, the Germanic system, as it developed after the Enlightenment of the eighteenth century, was less ideological but more abstract than the French Code. Whilst being centred on Germany, Austria and Switzerland, the Germanic Code influenced the development of legal systems throughout much of Central and Eastern Europe (prior to the area being absorbed by Communism in the mid-twentieth century); and although it might have been expected that Greece, on securing her inde- pendence from the Ottoman Empire in the early nineteenth century, would have adopted the Napoleonic Code, Greek law became modelled on the law of Roman Byzantium – the Greek legal system thereby establishing an affinity with the Germanic system, a relationship confirmed in 1946 when the Greek Civic Code adopted the Germanic legal style (Newman and Thornley, 1996). Under written constitutions, the Germanic system is Federal. In each country, power is shared between the central state and the regions (*länder*), and there is a variety of arrangements between the *länder* and the counties ('*kreise*') and communes (*gemeinden*). A number of free-standing cities in Germany (for example, Bremen and Hamburg) are, for historic reasons, vested with the combined powers and responsibilities of regional and local government, but in the former Austro-Hungarian Empire representative regional government is still comparatively weak – regional administrations formerly being little more than agencies of central government. In Greece, however, whereas the legal system is Germanic, administrative structures are distinctly Napoleonic.

In the Scandinavian countries of Denmark, Norway, Sweden and Finland, a hybrid version of Nordic and Germanic law emerged as a result of economic and cultural links with the states of north Germany in medieval times – an amalgam subsequently centralised and codified in the sixteenth and seventeenth centuries. In the early nineteenth century, and in the wake of the French Revolution, the legal style of Scandinavia was modernised and developed its own pragmatic course – eschewing the formalised Germanic codification of law. Similarly, hybrid administrative systems have emerged whereby, on the one hand, through a system of regional agencies (comprised of personnel appointed by the centre) there is a strong Napoleonic relationship between national government and the regions, while, on the other hand, autonomous local government – 'one of the cornerstones of the Scandinavian constitution' (Newman and Thornley, 1996: 35) – is a reflection of widely dispersed population and the strength of agrarian politics.

In the former Eastern bloc, the legal and administrative systems of the Czech Republic, Poland and Hungary are still in the process of transition, and there is uncertainty whether their common historic roots with Austria or Germany will re-emerge. Although in East Central Europe the restitution of property from the state to its former owners was the first stage in the creation of a market, attempts to market land were generally constrained by a deficiency of skills in property valuation and the absence of market-responsive planning (Newman and Thornley, 1996). There was also a slow and fraught process of decentralising decision-making and planning from the state to lower tiers of government. In Poland, decentralisation occurred even before the fall of the Communist regime in 1989. An Act of 1983 conferred explicit decision-making responsibilities on local authorities, not least within the field of local economic planning, although powers were constrained by Communist Party structures (Newman and Thornley, 1996). Although local authorities were freed from central political control by the constitutional reforms of 1989, their responsibilities and policies are still substantially under the control of central government due to economic and financial pressure (see Swianiewicz, 1992; Ciechocinska, 1994). Local authorities are thus faced with the difficulty of finding funds, providing new services such as housing and health services (formerly the responsibility of workplaces and professional organisations) and establishing expertise. Whereas reform, post-1979, has focused on the role of the 2500 communes, attention might soon shift to the middle-tier of government, the forty-nine regions (or voivodships) which act as central state agencies supplemented by a consultative assembly of people elected by the communes (Marcou and Verebelyi, 1993). There is possibly a need, instead, for about twelve regions with economic policy responsibilities, together with another tier of government between these regions and the communes to facilitate strategic planning. In the former Czechoslovakia, in contrast, the transfer of various responsibilities from the state to local government in 1990 was fairly straightforward, although its effects produced similar causes for concern. Reform necessitated the abolition of the Regional National Committees (which covered the ten provinces of the country) and the granting of self-government to 5769 communes (40 per cent of which were newly created). However, the resulting fragmentation at the lower level of government (the population of each commune averaged only 1800 inhabitants) made strategic planning virtually impossible and impeded investment in infrastructure and housing, and the provision of social services. Undoubtedly, pressures will mount for larger units of representative government – possibly at the level of region or sub-region.

Sub-national government in Western and Central Europe

With the disintegration of Christendom in the sixteenth century, the classic unitary state emerged initially in England, Spain and France, spread throughout parts of Western Europe in the seventeenth and eighteenth century, was created in Italy in 1870 and in Germany in 1871, and eventually became established in Central Europe with the collapse of the Austro-Hungarian and German empires and Imperial Russia after the First World War. Yet the development of the unitary state has taken several different forms. Currently its legal and administrative structures are either centralised or decentralised, but in a few instances, has recently become regionalised, whilst other countries have adopted a Federal structure with a degree of power-sharing between the central state and the regional-tier of government. A basic typology of states within the European Union, derived from Stoker, Hogwood and Bullmann (1995: 57) and Bullmann (1997: 5), but including a number of non-EU states, is set out in Table 3.1.

In 'classic unitary states', sub-national government exists only at the local level, but where regional structures occur these have been established for administrative purposes only and are strictly subordinate to central government. In terms of planning, this group of states can be sub-divided between unitary states with centralised planning powers (notably Luxembourg, Greece, Ireland, and the United Kingdom), and unitary states with planning powers substantially devolved to the municipalities (specifically Denmark, Norway, Sweden and Finland).

In respect of 'unitary states devolving power to the regions', elected regional authorities have been established as a consequence of governmental reform, the regional tier being guaranteed a degree of constitutional protection and autonomy. Notably in the case of Scotland, Wales and Northern Ireland, and in Portugal, France and the Netherlands, planning powers are

Table 3.1 A typology of regional government in Western and Central Europe

Classic unitary states	Unitary states devolving power to local authorities	Unitary states devolving power to the regions	Regionlised unitary states	Federal states
Luxembourg	Denmark	Portugal	Italy	Switzerland
Greece	Norway	United Kingdom	Spain	Austria
Ireland	Sweden	France		Germany
United Kingdom[1]	Finland[2]	The Netherlands		Belgium
		Finland		
States in political and economic transition				
Czech Republic ———————————————————→				
Hungary ———————————————————→				
Poland ———————————————————→				

Notes
[1] The United Kingdom was a classic unitary state until powers were devolved to Wales, Scotland and Northern Ireland in 1999 whence it became a devolving unitary state.
[2] Finland is in the process of devolving powers to some of its regions.

currently being (or have been) devolved to the regions – a process which could also be replicated in England and Finland in the near future, reducing their status as classic unitary states.

With regard to 'regionalised unitary states', directly-elected tiers of government – with constitutional status, wide-ranging autonomy and legislative powers – have been established in Italy and Spain, countries in which regional devolution has occurred to a greater extent than in any other West European unitary state – not least in respect of planning where power has been substantially devolved to the regions.

Federal states are distinguishable from unitary states since power and the co-existence of sovereignties is, in general, shared between the upper and regional tier of government; whereas the regional tier – existing in its own right – cannot be restructured unilaterally or abolished by the central government. However, in the Federal states of Austria, Switzerland, Germany and Belgium, despite power-sharing across a range of governmental functions, planning power, to a greater or lesser extent, is largely vested in the regions. 'States in political and economic transition' had much in common with classic unitary states prior to the collapse of Communism in the late 1980s since power was highly centralised. Under successive totalitarian regimes there was little or no devolution of power to representative sub-national government, but since the free elections of the late 1980s to early 1990s, devolution to the regions or even the adoption of Federal structures is becoming increasingly likely (see Chapter 9).

Whereas Figure 3.1 shows that spatial relationship between the five distinct legal and administrative systems in Europe and the sub-national levels of government in the EU, Table 3.2 demonstrates in more detail the very great differences in the degree to which regional government has emerged throughout the Member States of the Union. There are clearly substantial variations in the constitutional position of regional authorities with regard to electoral accountability, administration and planning responsibility, the right to participate in national policy-making, and the extent to which the regions exercise political and legislative control over sub-regional authorities.

The classification of sub-national territorial divisions

To facilitate a rational distribution of the three Structural Funds (the ERDF, the ESF and the guidance section of the EAGGF), the Statistical Office of the European Communities devised a common classification of sub-national territorial divisions for each Member State of the EU. Each country is consequently divided into *Nomenclature des Unités Territoriales* or NUTS based, as far as possible, on existing general purpose units (see Table 3.3). Among the 'classic unitary states', the whole of both Luxembourg and Ireland counts as a single region at NUTS-1 level, as well as at NUTS-2 level; whereas in Greece, in contrast, there are four NUTS-1 and thirteen NUTS-2 units. Of the 'unitary states devolving power to the local authorities', both Denmark and Sweden count as a single NUTS-1 unit, while the whole of the former country is also a NUTS-2 – in contrast to eight NUTS-2 in Sweden. Mainland Finland is also a single NUTS-1 unit, though a second Finnish NUTS-1 unit comprises the island of Ahvenanmaa/Aland. 'Unitary states devolving power to the regions' exhibit a diversity of nomenclature, in part reflecting their territorial size. Whereas in the United Kingdom there are two constituent countries (Scotland and Wales) and nine standard regions at NUTS-1 level, and in France there are nine NUTS-1 units, in contrast there are only four NUTS-1 units in the Netherlands, while in Portugal there are only three at this level (two of which are the autonomous island regions of the Açores and Madeira). In total these countries contained eighty NUTS-2 units. The two 'regionalised unitary states' of Italy and Spain have, proportionate to their populations, very broadly the same classification of sub-national territorial divisions: respectively eleven and seven NUTS-1 units, and twenty and eighteen units at NUTS-2 level. This is also the situation in the case of the 'federal states' of Austria and Belgium, which contained three NUTS-1 units and nine NUTS-2 units each, while in Germany there are sixteen NUTS-1 and forty NUTS-2, reflecting the country's much larger territorial size.

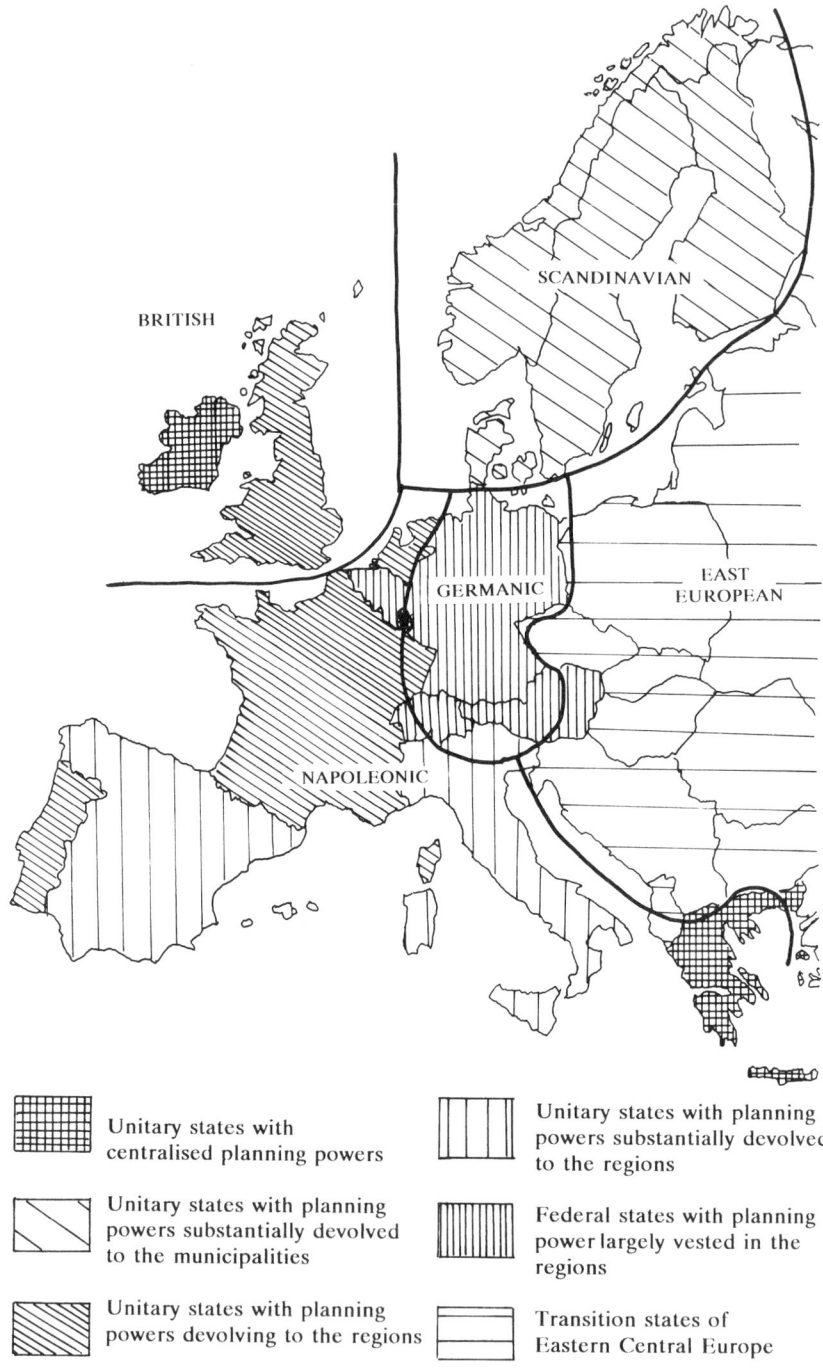

Figure 3.1 Types of states and government and planning systems
Source: Derived from Newman and Thornley, 1996, p. 29

Table 3.2 Sub-national constitutional differences, European Union, 1997

	Political regions [a]	Administrative/ planning regions [b]	Rights of region to participate in national policy-making	Political/ legislative control over sub-regional authorities
Classic unitary states				
Luxembourg	-	-	-	-
Greece	-	Development regions [13]	No	No
Ireland	-	Planning regions [9]	No	No
Unitary states devolving power to local authorities				
Denmark	-	Groups of Amter[3]	No	No
Sweden	-	Regional administrative bodies[8]	-	No
Finland	-	Regional councils[6]	-	No
Unitary states devolving power to the regions				
Portugal[c]	Island Regions[2]	-	No	No
United Kingdom	Scotland & Wales[22]	Standard regions[n]	No	Yes
France	Régions[22]	-	Consultative	No
Netherlands	Rijnmond Region[d]	Landsdelen[4]	No	Yes (in case of Rijnmond)
Regionalised unitary states				
Italy	Regioni[20]	-	No	Yes
Spain	Communidades Autonomas[17]	-	Yes	Yes
Federal states				
Austria	*Länder*[9]	-	Yes	Yes (not absolute)
Germany	*Länder*[16]	-	Yes	Yes (not absolute)
Belgium[e]	Communities[3]	-	Yes	No
	Régions[3]		Yes	Yes (not absolute)

Source: Adapted from Loughlin (1997)

Table 3.2 continued

Notes:
[a] Regions with an executive accountable to a directly elected assembly.
[b] Regions with administrative and planning responsibility, but accountable only to central government.
[c] Although the Portuguese constitution has conferred autonomy to the islands groups of the Azores and Madeira, government on the mainland remains highly centralised.
[d] It was decided in 1991 to set up a metropolitan region with an elected assembly in the Rotterdam area to replace the gemeente of Rotterdam and Province of South Holland. By 1997, this had not come into operation.
[e] By 1997, whereas the Flemish Linguistic Community and Flemish Economic Region had decided to form one body, the Walloon Community and Region remained as separate entities.

Sub-national territorial divisions and economic disparity

Within the fifteen Member States of the EU, the 77 NUTS-1 units, 206 NUTS-2 regions and 1,123 NUTS-3 units vary considerably in relation to living standards (as measured by their gross domestic products per capita), the proportions of their workforce employed in agriculture, industry and services, and their levels of unemployment. Comparing NUTS-1 units alone will broadly demonstrate these economic differences. Table 3.4 reveals that whereas a majority of units in classic unitary states, unitary states (with powers devolving to the regions) and regionalised unitary states had GDPs per capita *below* the EU average, the overwhelming majority of NUTS-1 in Federal states had GDPs per capita *above* the EU average, while one-half of NUTS-1 units in unitary states (with powers devolving to the local authorities) shared a broadly similar standard of living to Federal units. With regard to employment, the majority of NUTS-1 units in Federal states had a proportionately larger workforce in industry and smaller workforce in agriculture than any other group of units in the EU, whereas the opposite was the case in the NUTS-1 units of decentralised unitary states. The decentralised unitary states also contained the highest proportion of NUTS-1 with unemployment above the EU average, in contrast to unitary states devolving powers to the local authorities which had a small minority of NUTS-1 with below average unemployment rates.

From Table 3.3 and the commentary (p. 55) it can be inferred that, in general, the Federal states of Austria, Germany and Belgium have a higher standard of living than other states that are still in the process of devolving powers or remaining centralised, have a higher proportion of their workforce employed in industry and a lower proportion employed in agriculture and services than most other group of states, and have a moderately low level (but not the lowest level) of unemployment. In contrast, large parts of the regionalised unitary states of Italy and Spain have some way to go to raising their standards of living to Federal levels, reducing dependency on agricultural employment and cutting rates of unemployment. Clearly, classic unitary states (such as Ireland and Greece) and those

Table 3.3 National administrative divisions and NUTS Levels, EU, 1997

	NUTS 1		NUTS 2		NUTS 3	
Classic unitary states						
Luxembourg		1		1		1
Greece	Groups of development regions	4	Development regions	13	Nomoi	51
Ireland		1		1	Planning regions	9
Unitary states devolving power to local authorities						
Denmark		1		1	Amter	15
Sweden		1		8	Counties	24
Finland		2		6		19
Unitary states devolving power to the regions						
Portugal	Continente	1	Commissaoes de coodenaçao regional	5	Grupos de Cancelhos	30
	Regioes autonomas	2	Regioes autonomas	2		
United Kingdom	Scotland & Wales	2	Groups of counties	35	Counties/Local authority areas	65
France	SEAT	8	Regions	22	Départements	96
	DOM	1		4		4
Netherlands	Landsdelen	4	Provinces	12	COROP -Regios	40
Regionalised unitary states						
Italy	Gruppi di regioni	11	Regioni	20	Provencie	9
Spain	Agrupacion de communidades autonomas	7	Communidades autonomas	17	Prinvincias	50
			Melilla Y Ceuta	2		
Federal states						
Austria	Provinces	3	Länder	9	Regierungsberzirke	36
Germany	Länder	16	Regierungsbezirke	40	Kreise	543
Belgium	Régions	3	Provinces	9	Arondissements	43
EU 15		77		206		1123

Source: Adapted from Commission of the EC (1994): 173

Table 3.4 Gross domestic product per capita, employment and unemployment, NUTS-1, 1994–95

	Member State	Gross domestic product per head (PPS)[1] (EUR 15 = 100) 1994	Employment 1995			Unemployment rate
			Agriculture	industry	services	% 1995
NUTS-1 in classic unitary states						
Luxembourg	L	169	3.8	25.1	70.5	2.7
Ireland	IRL	88	12.0	27.7	60.0	14.3
Attiki	GR	73	1.1	26.9	72.0	11.0
Nisia Aigaiou, Kriti	GR	67	26.9	16.9	56.2	4.5
Voreia Ellade	GR	62	28.9	32.2	47.9	9.2
Kentriki Ellade	GR	57	38.2	19.5	42.3	7.4
NUTS-1 in unitary states devolving power to local authorities						
Ahvenanmaa/ Aland	SF	126	9.6	20.7	69.7	6.2
Denmark	DK	114	4.4	27.0	68.4	7.1
Sweden	S	98	3.3	25.8	70.9	9.1
Manner-Suomi	SF	91	7.7	27.6	64.6	18.2
NUTS-1 in unitary states devolving power to the regions						
Ile de France	F	161	0.5	20.8	78.5	10.0
South East	UK	117	0.9	21.9	76.7	8.6
West-Nederland	NL	113	2.8	18.6	75.8	7.3
Centre East	F	102	5.2	30.8	64.0	10.2
Noord-Nederland	NL	102	4.8	25.5	66.7	8.9
Zuid-Nederland	NL	101	4.0	27.6	65.3	6.9
Est	F	100	3.0	35.1	61.8	8.7
East Anglia	UK	100	3.8	27.4	68.4	6.7
Bassin Parisien	F	98	6.2	31.3	62.5	11.9
Scotland	UK	98	2.7	26.1	70.6	8.8
South West	UK	95	3.7	25.8	70.3	7.6
Sud-Ouest	F	93	8.6	24.4	66.9	11.0
Oost-Nederland	NL	93	5.0	25.3	66.2	7.1
East Midlands	UK	93	2.4	35.3	62.1	7.8
Ouest	F	91	8.9	29.2	61.9	10.5
Mediterranee	F	91	4.8	19.1	76.1	14.7
West Midlands	UK	90	2.2	34.2	63.2	8.8
North West	UK	88	1.1	29.4	69.0	9.2

Nord-Pas de Calais	F	87	3.3	29.5	67.2	15.3
Yorkshire and Humberside	UK	87	2.2	30.0	67.5	9.1
North	UK	85	1.8	31.1	66.2	11.0
Wales	UK	81	3.6	30.0	65.6	8.7
Northern Ireland	UK	80	5.8	26.1	67.5	13.0
Continente	P	68	11.2	32.4	56.4	7.1
Madeira	P	52	13.8	31.6	54.7	4.6
Açores	P	58	21.0	23.3	55.6	7.8

NUTS-1 in regionalised unitary states

Lombardia	I	131	3.5	41.8	54.7	6.1
Emilia-Romagna	I	128	8.4	35.0	56.5	6.3
Nord Est	I	119	6.3	37.2	56.5	6.0
Lazio	I	119	4.6	20.5	74.8	12.8
Nord Ouest	I	116	4.8	36.9	58.3	8.7
Centro	I	107	5.5	35.6	58.9	8.0
Madrid	E	95	0.9	25.4	73.7	20.7
Noreste	E	89	6.6	37.2	56.2	19.3
Abuzzi-Molise	I	87	10.2	23.3	66.6	25.9
Este	E	86	4.7	36.5	58.8	20.3
Sardegna	I	78	15.3	24.6	60.1	20.8
Canarias	E	75	7.0	18.7	74.3	23.7
Sicilia	I	70	13.2	18.4	68.4	23.3
Campania	I	69	9.7	31.3	59.0	10.8
Sud	I	68	15.7	23.2	61.2	18.8
Centro	E	65	15.5	29.6	54.9	22.4
Noroeste	E	64	23.1	26.9	50.0	18.5
Sur	E	58	12.5	23.6	64.0	31.8

NUTS-1 in Federal states

Hamburg	D	196	1.1	24.9	74.1	7.6
Region Wallonne	B	183	2.8	25.2	72.0	12.9
Bremen	D	156	0.7	28.4	70.9	10.6
Hessen	D	152	2.4	33.8	63.8	6.3
Bayern	D	128	4.9	37.5	57.5	4.9
Baden-Württemberg	D	126	2.5	43.4	54.1	5.5
Östosterreich	A	122	6.1	28.7	65.2	
Bruxelles-Brussel	B	115	0.1	15.8	84.1	13.3
Nordrhein-Westfalen	D	112	1.9	36.0	62.1	8.2
Westosterreich	A	110	7.4	35.3	57.3	
Saarland	D	106	1.0	35.1	63.9	9.1
Schleswig-Holstein	D	106	4.4	28.9	66.7	6.5
Niedersachsen	D	105	4.4	34.0	61.5	7.9
Berlin	D	104	0.8	24.8	74.4	11.2
Rheinland-Pfalz	D	100	3.3	36.9	59.7	6.2
Vlaams Gewest	B	91	3.0	31.4	65.6	6.9
Sudosterreich	A	87	9.9	33.3	55.9	-

Table 3.4 continiued

	Member State	Gross domestic product per head (PPS)[1] (EUR 15 = 100) 1994	Employment 1995			Unemployment rate
			Agriculture	industry	services	% 1995
NUTS-1 *in Federal states (continued)*						
Sachsen	D	60	2.6	39.8	57.6	13.8
Sachsen-Anhalt	D	60	4.8	37.2	57.9	16.7
Thüringen	D	60	3.6	37.7	58.7	11.9
Mecklenburg-Vorpommem	D	57	7.1	29.6	63.3	11.9
EUR 15		100	5.3	30.2	64.3	10.7

Source: Office for National Statistics (1997) *Regional Trends, 1997*, The Stationery Office, London

Note: [1] Purchasing Power Standard

already in the process of devolving their powers (and especially their planning powers) to regional authorities (for example, the United Kingdom, France and the Netherlands) will need to ensure that constitutional change towards more Federal structures takes place in a manner compatible with economic growth and a more spatially-equitable distribution of GDP per capita and employment opportunities. The same would be desirable if the East Central European states of the Czech Republic, Hungary and Poland chose to embark upon a process of devolution to a regional-tier of representative government.

4 Unitary states with centralised planning powers

With devolution occurring in the United Kingdom, it is evident that the most unitary states in Western Europe are Luxembourg, Ireland and Greece. Luxembourg is a unitary state almost by default since it is too small to have regions in the sense in which they are recognised within the EU. The whole country is deemed to be a single region for the purpose of Structural Fund allocations. Ireland is also a region for EU funding, but its basic political and administrative units are its central government and the counties (dominated by non-elected officials). Centralised power is a legacy of the 'former colonial administration based in Dublin' and the 'centralising tendencies of Irish nationalism' (Jeffrey, 1997: 152). In response to serious criticism, there has been some attempt to decentralise administration with the establishment of nine Regional Development Organisations (RDO) in 1994, but this has done little to reduce the dominance of the state/county system. The Greek system of government is modelled on the Napoleonic system of 'a highly centralised state and prefectorial control of sub-national entities' (Jeffrey, 1997: 152), which has survived despite recent attempts to decentralise power and to establish a regional level of administration.

Luxembourg

Due to its very small size (2,586 km^2), the Grand Duchy of Luxembourg is inevitably a unitary state and has little opportunity to develop a regional tier of administration. There are nevertheless three tiers of local government: three districts, twelve cantons and 118 communes. However, under the Planning Act of 1974, spatial planning operates at only two levels: at the national level and at the level of the commune (EC, 1994).

At the instigation of the minister responsible for spatial planning, the Council of State produces a master plan (*programme directeur*) which sets out the main objectives and the measures necessary to achieve them over a period of ten to twenty years. Although the plan is not binding on third parties, it nevertheless needs to be taken into account as an instrument of co-ordination. Because the 1988 version of the *programme directeur* was designed to expire in the year 2000, the Ministry of Planning began work on preparing a new master plan in 1995 for the first two decades of the twenty-first century.

General and local development plans are also drawn up by the Council of State. As vehicles for implementing the master plan, they are declared legally binding subject to specific participation and consultation procedures. By the mid-1990s, they covered such developments as the creation of national industrial areas, the surroundings of the airport and land reclamation.

At the lower tier of government, the communes are responsible for their own planning – provided that they are equipped with planning laws which cover their entire area. Under the 1974 Act, communal development plans must conform to general and local development plans, must take account of individual objections, and finally must be submitted to the centrally-appointed Planning Commission.

Regional policy and regional incentives

Nationally, economic growth in Luxembourg is constrained by the need to restructure the old and declining coal and steel industries of the south and to diversify and strengthen the 'fragile' agricultural areas of the north. Thus, under an Industrial Framework Law of 1986, and throughout most of Luxembourg, incentives became available to small and medium-sized enterprises (SMEs) in the form of capital grants, interest rate subsidies, loan guarantees and tax concessions. In 1993, however, new legislation limited assisted-investment to six designated cantons: Grevenmacher, Luxembourg, Esch-sur-Alzette and Capellen in the south, and Clervaux and Wiltz in the north, where maximum rates of assisted-investment would range from 17.5 to 25 per cent of eligible expenditure compared to 7.5 to 15 per cent elsewhere, and support for small and medium enterprises, research and development, energy saving and environmental protection was enhanced.

Since there was an absence of areas in Luxembourg with development lagging behind, there were no Structural Fund Objective 1 allocations for the period 1994–99. However, Objective 2 funding amounting to 15 billion ECU was targeted at converting the economy of the coal and steel areas of the south within the cantons of Esch-sur-Alzette and Capellen, while Objective 5b funding of 6 billion ECU was directed at diversifying the rural areas of the north, particularly in the cantons of Clervaux and Wiltz (Figure 4.1). Since the whole of Luxembourg is simultaneously a NUTS-1, NUTS-2 and NUTS-3, central government alone was involved in seeking funding from the European Commission. As many as 42 per cent of the population of Luxembourg lived in areas eligible for Structural Funding – 34.2 per cent in Objective 2 areas and 7.8 per cent in Objective 5b areas (CEC, 1996b).

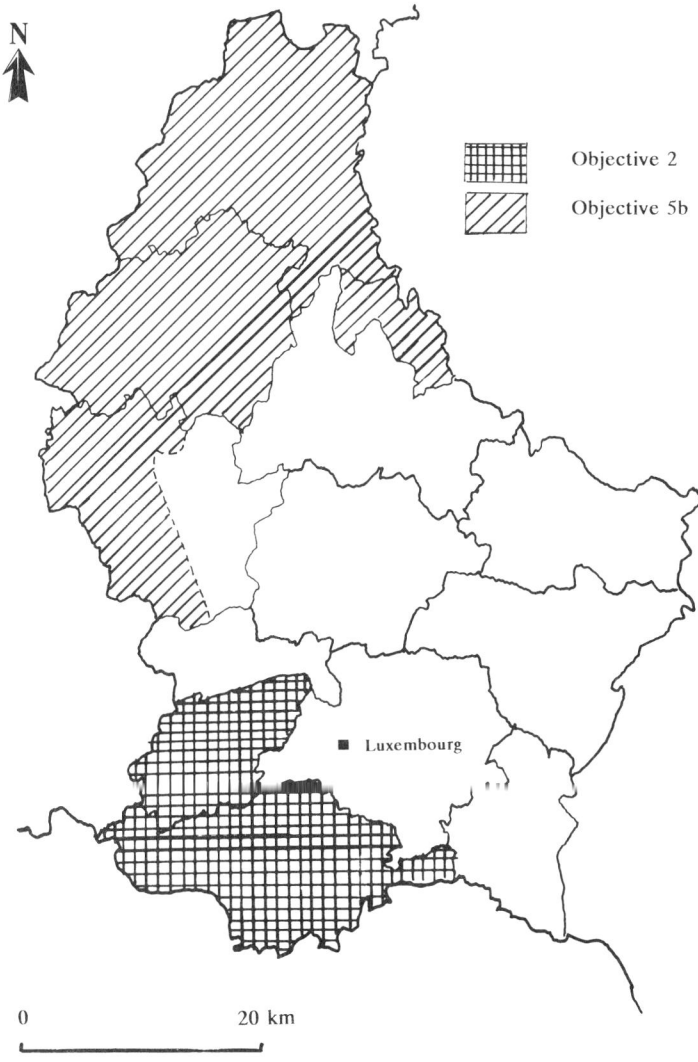

Figure 4.1 Designated Structural Fund areas, Luxembourg, 1994–99

Ireland

Although Ireland is considerably larger than Luxembourg (it has an area of 68,893 km^2), like the Grand Duchy it has a highly centralised system of government – a legacy of British colonial rule and the predeliction of Irish nationalism which has militated against the development of regionalist and federalist traditions (Jeffrey, 1997; Loughlin, 1992).

Under the Local Government (Planning and Development) Act of 1963 and subsequent amendments, the central government is responsible for formulating planning regulations which establish the procedures for preparing and reviewing development plans and for dealing with applications for planning permission. In addition, central government publishes policy guidance statements and, where appropriate, can insist that development plans are amended. It also produced the National Development Plan 1994–99 which sets out its medium-term development strategy in relation to the national and Community objective of greater economic and social cohesion (EC, 1994). The Plan is strongly related to the development of opportunities eligible for EU funding rather than an attempt to comprehensively assess national priorities (Newman and Thornley, 1996), and in its spatial and economic manifestations has an impact on major infrastructure projects.

To avoid jeopardising the allocation of EU funds to the whole country as a designated single region, the Irish government has eschewed the creation of regional administrations and the development of inter-regional policy. However, there is a regional tier of nine authorities established under the Local Government Act of 1991 (Figure 4.2), with each regional council being composed of members appointed by the local authorities in the region. Since January 1994, the regional councils have attempted to co-ordinate the provision of public services in their region and integrate the policies of central government ministries at a local level, and have reviewed the development plans of local authorities with regard to consistency and the implementation of EU funded programmes (Newman and Thornley, 1996).

At a local level, a total of eighty-eight counties and urban authorities have a statutory duty to produce development plans indicating development objectives for their areas. Development plans do not have to be approved by a higher authority, but all development must be in accordance with the development plan and the relevant local planning requirements, and be subject to planning permission granted by the local authority (Newman and Thornley, 1996).

Regional policy and regional incentives

Although Ireland as a whole has a disproportionately large agricultural workforce and an underdeveloped urban hierarchy, and despite the whole of the country qualifying for aid under EU Regional Policy, the western counties are markedly poorer and more marginal than the east (Minshull, 1996).

Current incentive policies originate from the establishment of the Irish Development Authority in 1949 which ended restrictions on the foreign ownership of Irish companies and encouraged industrial investment particularly in export-orientated industries. Subsequently, industries in the twelve western counties of Ireland were eligible for a higher rate of aid than in the rest of the country under the provisions of the Underdeveloped Areas Act of 1952, and the 'Small Industries Act' of 1969 aimed to assist

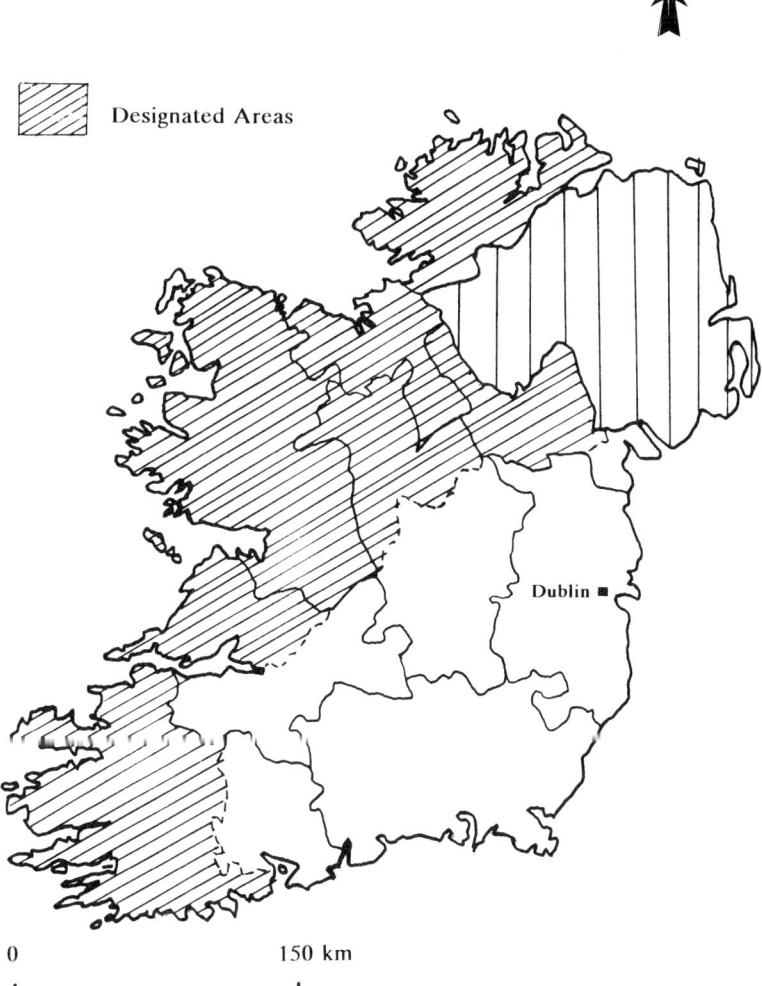

N

Designated Areas

Dublin ■

0   150 km

Figure 4.2 Government assisted regions, Ireland

the modernisation and enlargement of local craft industries widely distributed throughout the west (Minshull, 1996). In the 1970s and 1980s, the
Irish Development Authority attracted investment into the three regional
centres of Cork, Waterford and Limerick, and extended incentives to
Dublin in 1983.

In the 1980s, as part of a programme to rein back public spending, aid
became increasingly more selective and aimed to enhance competitive

ability. Under legislation in 1986, there was a shift of emphasis from assisting job creation per se towards the objective of maximising added value to the national economy, and in 1987 support became more target-orientated and performance-related than hitherto. In 1988, support for fixed asset investment was reduced, with the aid ceiling for extensions to plant being reduced from 60 or 45 per cent of eligible expenditure (depending on location) to a standard 25 per cent, while support to small firms was henceforth targeted at the development of firms with growth potential rather than at start-ups in general (Yuill *et al.*, 1996). Although the designated areas were increased in size in 1989 (and thereby increasing their share of the national population from 28 to 34 per cent), the rate ceiling for extensions was further reduced from 25 to 15 per cent in 1991.

In the early 1990s, it was recognised that the promulgation of indigenous industry was crucial for the development of self-sustaining Irish economy, and that a reform to the system of regional aid was necessary. In 1993, the Industrial Development Authority was therefore replaced by *Forfas*, which, through the medium of its sub-agency *Forbairt*, had the responsibility of more effectively promoting and funding the development of indigenous industry, and, through a newly constituted Industrial Development Agency – Ireland (IDA–Ireland), of encouraging more inward investment from overseas (Minshull, 1996). *Forbairt* soon introduced a Small Industry Programme comprising a discretionary package of capital or employment grants, mentor grants, grants for research and development and grants for training. IDA–Ireland also awarded capital grants, employment grants and grants for research and development and training, but in addition provided interest relief grants, loan guarantees and rent subsidies. In the Designated Areas the maximum rate of award was set at 60 per cent of eligible expenditure compared to 45 per cent elsewhere (Yuill *et al.*, 1996).

Although the regional incentive policy applied by central government over the years was in part responsible for Ireland's GDP per capita increasing from 68 per cent of the EC average in 1984 to 88 per cent of the EU average in 1994, Ireland, in the mid-1990s, still suffered from high unemployment, a comparatively low level of industrialisation (characterised on the one hand by a largely foreign-owned, highly productive, export-orientated and modern sector, and on the other hand by a traditional, labour-intensive and indigenous sector), a concentration of population on the eastern seaboard with a small and dispersed population in the west, and a high level of peripherality and poor communication infrastructure in relation to major European markets.

Since, in economic terms, the whole of Ireland was clearly lagging behind most other EU countries, it became eligible for Objective 1 funding at NUTS-2 level – receiving an allocation of 5.6 billion ECU, 1994–99 (CEC, 1996b). Comprising a detailed development strategy and a justification for the Community Support Framework, Ireland's *National Development Plan*

1994–99 provided a framework for the implementation of Operational Programmes concerned with the following priorities (Williams, 1996):

1 The development of the productive sector, including agriculture, forestry, fishing, manufacture, food-processing and tourism. An allocation of 38.9 per cent of the budget.
2 The development of human resources and education and the promotion of social integration. An allocation of 37.6 per cent.
3 The development of a modern economic and communication infrastructure, energy production and transmission, and enhancement of the environment. An allocation of 19.8 per cent.
4 Assistance to local enterprises in urban and rural areas. An allocation of 3.6 per cent.

If successful, these programmes will ensure that Ireland's GDP per capita will further converge towards the EU's average. Beyond 1999, further development will increasingly take place within an all-Ireland context. As an outcome of Anglo-Irish talks (1993–98), and under a North-South Ministerial Council, representatives from the Dublin government and a new Northern Ireland assembly will share responsibility for matters of common concern such as urban and rural development, the environment, transport planning, tourism and relations with the EU.

Greece

The Greek administrative approach is highly centralised – a legacy of the country adopting the French prefectural system in the 1830s. Although there are various sub-national tiers of administration – 13 *periferies* (regions), 54 *nomoi* (districts), 264 *demoi* (cities of more than 10,000 inhabitants) and 5,759 *kinotites* (communes of more than 1000 inhabitants) – only the two lower levels are elected. Responsibility for planning resides, to a varying extent, with central government, the regions, the cities and the communes, but, because of the complexity of the planning system and problems of enforcement, it could be argued convincingly that there has been a 'long standing coalition acting against planning involving the state; landowners, developers and certain professions' (Delladetsima and Leontidou, 1995).

Although central government in the 1990s has been reluctant to produce a national spatial plan despite legislation authorising planning at the national level, it nevertheless presides over a highly centralised planning systems and is involved in planning policy at all levels of administration, including development control matters at the level of the city or commune (Delladetsima and Leontidou, 1995; Newman and Thornley, 1996). Even building permits are administered by local offices of central government.

Non-statutory regional development plans are, however, the responsibility of the regions – legal entities created for the planning and co-ordination of regional development but without any degree of autonomy. 'They are devolved entities of central government without an administration of their own and are headed by a secretary general, who is appointed by the central government and presides over a regional council' (Wiehler and Stumm, 1995: 238). Introduced in 1980, regional plans have been formulated to guide the allocation of resources facilitated by special national programmes, but are also used in negotiations for EU funding. Consequential public sector development often has a marked effect on spatial planning at lower levels of administration (Newman and Thornley, 1996).

At the level of the district, centralised control is similarly evident. Each ministry is represented at this level of administration and the chief admin-istrator – the prefect – is appointed by the Minister of the Interior. Districts or prefectures, however, have very few planning responsibilities and these are largely confined to approving the development plans of communes of minor importance.

The cities and communes alone have had responsibility for the prepara-tion of plans at a local level since 1923. But although city and communal authorities could draw up plans under the 1923 Law for either the whole of their area or just one block, from the 1920s plans had to be submitted to the central ministry for approval with the possibility that they would be modified or rejected (Newman and Thornley, 1996). Although still the basis of planning in some local authority areas, the 1923 Law provides no opportunity for planning within a regional context, partly ignores private land use and pre-dates the notion of participation.

Following the return of parliamentary democracy after the fall of the military dictatorship in 1974, the planning system was reformed by the 1979 Law but the centralised administrative structure remained largely intact. The 1979 Law required city and communal authorities to produce development plans under central supervision, and stated that these, rather than a codification of norms as before, should form the legal framework for private development (Getimis, 1992). In an attempt to ensure that afford-able land was available for infrastructure development and communal use, development plans were modelled on the French *zone d'aménagement concerté* principle that future development should take place only in designated 'urban development areas'. The full implementation of development plans, however, was soon impeded by private landowners who frequently objected to being prevented from realising the full development value of the land or even having to forgo their property.

In the 1980s, the first Socialist government in post-war Greece embarked upon a policy of decentralising and participation. It proposed that powers to supervise the appointment of local government officials, control develop-ment and regulate building should be devolved from central government to mayors and local authorities (Newman and Thornley, 1996). Under the 1983 Law, local authorities were given powers to formulate two levels of plans: a general development plan and implementation plan which had to be compatible with it, and the power to approve development plans was devolved to city and communal administrations. Strategic master plans were also introduced, and by the mid-1980s were being prepared by Athinai and Thessaloniki. In addition, there was a programme to introduce building regulations for 10,000 rural settlements and plans for 1800 developing settlements in rural areas (Makridou – Papdaki, 1992), while an Operation for Urban Reconstruction was launched to check hitherto uncontrolled urban sprawl. However, with the economic crisis in 1986 and the election of a Conservative government in 1990, the further development of a decentralised and participatory system of planning ceased (Newman and Thornley, 1996).

Over the years, Greek planning law has become increasingly complex and subject to frequent revision – largely because there is no political consensus over the aims and objectives of planning and methods of enforcement. Even in the mid-1990s, central government continued to involve itself at all levels of planning, and development control still remained a central government function (Delladetsima and Leontidou, 1995). Yet it was becoming evident that reforms were being seriously considered by the Minister for the Environment, Spatial Planning and Public Works. An elected tier of regional authorities with planning responsibilities was under discussion, and below the regional tier it was possible that prefects would soon be elected and district councils (*nomoi*) would have a degree of autonomy including the responsibility for making staff appointments.

Whether Greece will develop an effective system of spatial planning in the near future is, however, far from certain. Throughout most of the twentieth century, spatial planning has been 'undermined through the network of family, extended kinship and political ties, and generally exploited for personal and political gains' (Wassenhoven, 1984: 7). As long as planning is in a constant state of flux and enforcement powers are limited, inevitably there will be 'a considerable gap between official planning and the reality of development on the ground' CEC (1994c).

Regional policy and regional incentives

The topography and the peripheral location of Greece, together with its limited natural resources, renders Greece an economically poor country by European standards, its regional GDP per capita ranging from only 35.2 to 58 per cent of the EU average, 1989–91. With marked economic disparities between the more prosperous industrialised areas of Athinai and Thessaloniki, and the poorer agricultural regions, Greek regional policy has for a long time aimed to promote productive investment in the less developed regions and to curb development in Athinai and other expanding areas.

In 1981, Law 1116 introduced both fiscal and financial packages to stimulate regional development – the former measures comprising a tax allowance and an increased depreciation allowance, and the latter comprising an investment grant and interest rate subsidy. With a change from a conservative to a socialist administration in 1982, Law 1262 placed most emphasis on financial measures and introduced a decentralised approach to incentive administration and re-drew the assisted areas map. However, under a new conservative administration in 1990, Law 1892 increased the stress placed on the fiscal package but reduced the overall weight given to regional incentives – increasingly channelling aid instead through infrastructure spending derived from EC Structural Fund. New legislation was introduced in 1994 (Law 2234), increasing the weight attached to the financial package while maintaining the importance of fiscal measures. The administration of aid, however, was re-centralised.

Like Ireland, the whole country was eligible for Objective 1 funding at NUTS-2 level. A total sum of 14 billion ECU was made available over the period 1994–99 – topping-up the financial and fiscal package introduced under Law 2234 (CEC, 1996b).

5 Unitary states with planning powers substantially devolved to the municipalities

The Scandinavian nations, Denmark, Sweden, Norway and Finland, have, in almost identical ways, devolved their responsibilities for spatial planning to a greater extent than any other group of countries in Europe, largely as a result of a cross-fertilisation of ideas and policy (Hall, 1991). With central government retaining only minimal responsibility for planning, municipalities, rather than the regions or counties, have become the principle planning authorities. Similar to practice in Germany, Austria and Switzerland, broad general plans (indicating public sector activity and matters of national concern) form the framework for legally binding detailed plans, and are only applied when development is anticipated. Only at this stage is planning permission granted, an outcome of negotiation between the municipality and the developer (Newman and Thornley, 1996).

Denmark

Municipal self-government and decentralised decision-making are key principles of the Danish Constitution, although, in respect of the three-tier planning system and under the Planning Act of 1992, municipal plans must conform to county plans and these in turn must be compatible with the national plan.

The national government is, of course, responsible for matters of national importance. Following elections every four years, it prepares a national planning report (the *Landsplantedegegorrelse*), which in 1992 was issued as a spatial development perspective (the *Landsplanperspektiv*) entitled *Denmark towards the Year 2018*. It proposed that a network of cities should be developed within Greater København to help create (with the further development of Malmo in Sweden) one of the strongest regions in Europe But like other Danish national plans, it was not binding on lower levels of government (Östergård, 1994). National government, nevertheless, attempts to relate planning regulations to major land-use categories. The whole country is therefore divided into three zones: urban, recreational and rural, with the latter zone covering about 90 per cent of the country. In the urban and recreational zones, development has to conform to local plans and regulations, but in rural zones all development is prohibited except that concerned with commercial activity directly relating to agriculture, forestry and fishing (Newman and Thornley, 1996). It also regulates the way in which legislation at county and municipal level is implemented, it can issue national planning directives (*Landsplandirektiven*) to lower tiers on issues such as infrastructure and landscape protection, it can veto regional plans and it can call in municipal plans for consideration (Newman and Thornley, 1996).

At the middle level of government, the fourteen county councils (*amtskommuner*) and the cities of København and Frederiksberg are legally obliged to establish and adopt regional plans (*regionplaner*) for their areas. Regional plans need to comply with the regulations and directives issued by national government, and, like the national planning report, are subject to post-election revision every four years (Newman and Thornley, 1996). Covering the whole of a county and binding on municipal plans, a regional plan sets out policies on urban growth, urban zones, the location of major developments and environmental protection. At this level, the Danish environmental assessment procedure is incorporated into the planning system. Planning decisions in rural zones are also the responsibility of the county (European Commission, 1994; Newman and Thornley, 1996).

With average populations of about 20,000, the 275 municipalities are the principal authorities for planning in Denmark, and each approves its own plans. In accordance with regional guidelines, the municipality prepares a structure plan (*kommuneplan*) and a local plan (*lokalplan*). The former plan, reviewed every four years after municipal elections, comprehensively covers the whole area of the municipality, sets out the broad pattern of uses, indicates zonal change (for example, from rural to urban), and sets out a framework for local plans (Newman and Thornley, 1996). In the 1990s, structure plans have become less regulatory and more concerned with, for example, urban renewal, the environment *versus* commercial growth or political objectives. Local plans, in contrast, are more regulatory specifying land use and the size, location and appearance of buildings, and are legally binding on property owners. Local plans might promote part of the area of the municipality or control both large and small scale development, or there may be more emphasis on thematic issues such as neighbourhood enhancement or conservation of areas of historic interest. If a proposed development is in accordance with the local plan and local regulations, then planning permission is automatically granted, although development necessitates the acquisition of a building permit (*Byggetilladel*) (European Commission, 1994; Newman and Thornley, 1996).

At county and local levels, there is ample opportunity for public participation and debate at each stage in plan formulation process, and, at each level of planning, public opinion is again reflected in the preparation of revised plans every four years following elections (Newman and Thornley, 1996).

Regional policy and regional incentives

With the rate of unemployment in Denmark being comparatively low, 8.7 per cent in 1994, compared to an aggregate of 11.3 per cent in the European Union as a whole, and with only minor regional variations in unemployment within the country (the level of unemployment being highest in north-east Jylland and in Lolland), regional aid has diminished quite considerably in recent years. To bring this about, the boundaries of the General Development Regions were progressively tightened, reducing their population from 31 per cent of the national total in the late 1970s to 24 per cent in 1984; soft loans to companies and a range of minor incentives were withdrawn in 1985; there was a further reduction in the size of the problem regions in 1987, diminishing their population to 20 per cent of the national total; support became more selective in 1988 and was targeted at projects with significant development impact; and conventional regional development grants and loans were discontinued in 1991 in favour of a new business development system (CEC, 1994c).

Since there was an absence of any region in Denmark whose development was lagging behind, there were no Structural Fund Objective 1 allocations for the period 1994–99. However, Objective 2 funding, amounting to 119 million ECU for the period 1994–99, was targeted at eligible NUTS-3 (the *amtskommuner*) to convert the economy of areas seriously affected by industrial decline, and 54 million ECU were available to Objective 5b areas to facilitate the development and structural adjustment of the rural economy over the period 1994–99 (see Figure 5.1). Applicant *amtskommuner* were obliged to submit conversion and development plans to the central government for consideration as the first stage of seeking funding from the Commission. Altogether, areas containing 15.8 per cent of the Danish population were eligible for funds, 8.8 per cent in Objective 1 areas, and 7 per cent in Objective 5b areas (CEC, 1996b).

Sweden

For over thirty years after the Second World War, the Social Democrats maintained sizeable majorities in government and applied a planning system which ensured that the state had considerable powers to control development at national and local levels (Newman and Thornley, 1996). Under the Planning Act of 1947, central government was given the responsibility of producing a master plan for the whole country with more detailed plans for urbanised areas. The master plan, intended as a guidance document for the municipalities, indicated future land uses, it was reviewed every five years and from time to time was subject to ratification at national level. The pace, type and location of development, however, was also influenced by Sweden's massive social housing programme in

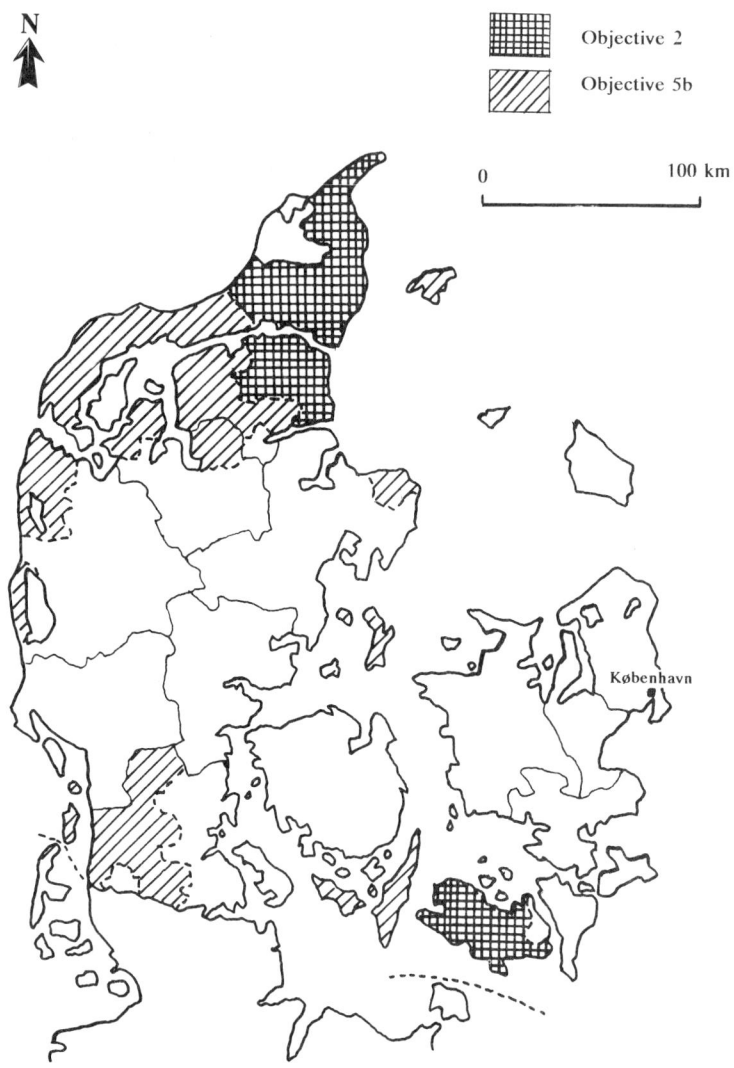

Figure 5.1 Designated Structural Fund areas, Denmark, 1994–99

the 1960s, the redevelopment of town centres, and by municipal land banking (Newman and Thornley, 1996; Duncan, 1985).

By the 1970s, it was increasingly recognised that the Swedish planning system was too inflexible, and soon the move away from state intervention in the economy to greater market freedom, experienced throughout most of Europe in the 1980s and 1990s, was accompanied in Sweden by the decentralisation of administration and political structures from the upper to the lower tier of government. Thus, in response to the rigidity and increasing irrelevance of centralised planning, and to the demand for increased public participation at the grassroots, reforming Acts in 1987 clarified the planning role of central government and conferred a substantial amount of responsibility for planning on the municipalities.

Under the Natural Resources Act of 1987, therefore, the planning role of central government was restricted to the locational control of power stations, chemical plants and other hazardous industries and to the imposition of planning guidelines on, for example, national recreation areas, ecologically sensitive areas and water supply.

As in other Scandinavian countries, the middle-level of planning is very weak. Whilst the twenty-four counties and larger metropolitan areas do prepare regional plans, these only provide a broad overview, are mainly statements of research and co-ordination, and are only advisory (Newman and Thornley, 1996). It can also be anticipated that larger regional units will emerge as the counties increasingly co-operate in the preparation of regional plans.

At the lower level of planning, the role of the 284 municipalities has been enormously strengthened by the Planning and Building Act of 1987. Under the legislation, all municipalities are obliged to produce a comprehensive plan (*översiktsplan*) indicating, for their whole area, the intended pattern of land use and development. Whilst not binding on individuals, and allowing a great deal of flexibility, comprehensive plans provide the framework for the provision of public services and for more detailed planning. Although municipalities enjoy a great deal of autonomy from county and central government, comprehensive plans are scrutinised by the County Administrative Board, an agent of central government, which checks whether they have adequately taken national interests and health and safety into account. Comprehensive plans are also the subject of consultation with the public and other interested bodies, and municipalities are required to demonstrate how they respond to the views articulated.

More important than the comprehensive plan, the legally binding detailed plan (*detaljplan*), prepared only when development is anticipated, specifies intended land uses, public uses, building lots, design, construction materials, floor areas, landscaping, parking, conservation and the implementation period (Kalbro and Mattsson, 1995; Newman and Thornley, 1996). During the preparation stage of the plan, municipal planners are obliged to consult with other public bodies, developers and other interested parties and to facilitate public participation, prior to the plan's ratification by the municipality. To ensure that the detailed plan conforms with the national interest it is subject to scrutiny by the County Administrative Board to whom appeals can be made, which if unsuccessful, can then be made to the central government. If the proposal conforms to the ratified plan, the developer will seek a building permit, which is normally granted automatically (Newman and Thornley, 1996). Under the Special Area

Regulations of the 1987 Act, however, some comprehensive plans are also legally binding if they aim at protecting the national interest (for example, in the case of areas for holiday homes), but they do not, in contrast to detailed plans, confer development rights (Newman and Thornley, 1996). In these circumstances, development proposals must confirm with the detailed plan (as influenced by the legally binding comprehensive plan) and the developer would be required to seek a building permit in the normal way.

Regional policy and regional incentives

In contrast to the non-Nordic countries of Europe, regional policy in Sweden is less concerned with the need to reduce spatial disparities in output and employment, than with the aim of maintaining population, employment and incomes in the more remote areas. A system of long-term transfers was introduced to partly compensate for difficult natural living and working conditions, low population densities and long travel distances. Since Sweden contains a relatively well developed infrastructure, regional incentives are directed at business support and public services rather than at transport and communication investment, and take the form of development and investment grants, loans and guarantees, and transport user subsidies. In many cases, incentives are regionally differentiated to cross-subsidise the costs of transportation to and from the less favoured regions (CEC, 1994). With the move towards a freer market economy, however, regional incentives emanating from central government were increasingly at risk of being eroded during the latter years of the twentieth century.

Notwithstanding the very real need for regional incentives, there are in fact no regions in Sweden whose development is lagging noticeably behind the rest of the country, and therefore there is an absence of Objective 1 funding. However, as is shown in Figure 5.2, the relevant counties and metropolitan areas, the NUTS-3, were eligible for Objective 2 funds amounting to 157 million ECU, 1994–99, to convert the economies of areas of industrial decline. In addition, a further 135 million ECU were available under Objective 5b funding to assist the development and structural adjustment of rural areas, and Objective 6 regions (the northern counties of Norrbotten, Västerbotten and Jämtland) became eligible for funds amounting to 247 million ECU, 1996–99, to facilitate development and structural adjustment in areas of extremely low population density (CEC, 1996). Counties and metropolitan areas applying for funds were obliged to submit conversion and development plans to the central government for consideration as a preliminary to seeking funding from the Commission. Only 24.6 per cent of the Swedish population, however, lived in regions eligible for Sructural Funds, 11 per cent in Objective 2 areas, 8.6 per cent in Objective 5b areas and only 5 per cent in the extensive Objective 6 areas (CEC, 1996b).

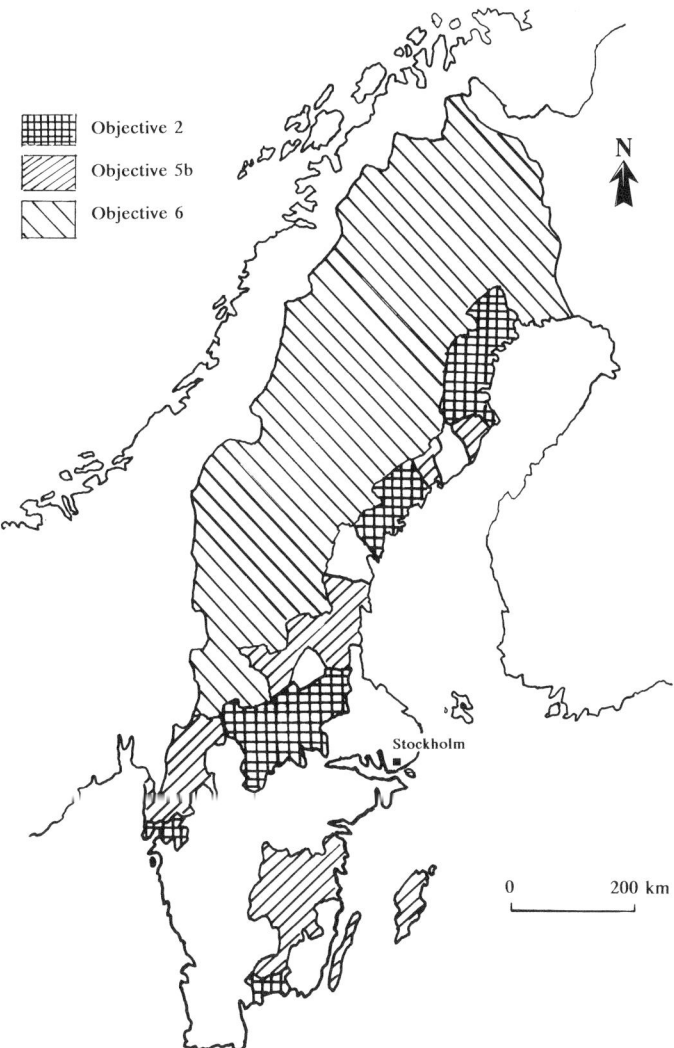

Legend:

- Objective 2
- Objective 5b
- Objective 6

N

0   200 km

Stockholm

Figure 5.2 Designated Structural Fund areas, Sweden, 1995–99

Norway

Since the 1960s, the move to decentralise decision-making from the national state to the nineteen counties and 454 municipalities has given local government, and particularly the municipalities, considerable power (Lorange and Myhre, 1991). This undoubtedly was intended to satisfy the democratic aspirations of a sparse and highly dispersed population (see Falkanger, 1986).

Under the Planning Act of 1965 (with revisions included in the Planning and Building Act of 1986), a hierarchy of plans was established comprising national, regional, county, municipal and local plans. Although the national government was given responsibility for producing highway and telecommunication plans and regional strategies to protect peripheral areas, regional planning per se was very low key and unrepresentative of any political entity (Newman and Thornley, 1996). Under the 1965 Act, regional plans were often drawn up, with only a degree of success, to deal with inter-county issues such as water supply or sewage disposal, and were consequently abandoned in the 1970s (Newman and Thornley, 1996); and in the Oslo region (except for major infrastructure schemes) there is little regional control over the independent plans of over thirty municipalities (Lorange and Myhre, 1991).

At a county level, planning has been relatively more successful, partly because county authorities are democratically elected. Since 1973, obligatory and comprehensive county plans (*fylkesplan*) focus on those items for which they have a financial responsibility such as county roads and hospitals, but also provide guidelines for lower-tier planning in respect of such items as population growth and migration, the use of natural resources and economic development (Newman and Thornley, 1996).

Lower-tier planning, by municipalities, is the dominant form of planning in Norway. Since 1965, the 403 municipalities have been obliged to produce comprehensive and binding municipal plans (*Kommuneplaner*) to provide the framework for local plans. Essentially concerned with the land-use zoning throughout the municipality, local plans are subject to county and ministerial scrutiny before they can be applied. Under the 1986 Act, the approach to municipal planning was amended to place an emphasis on economic and social issues, the co-ordination of plans, and public participation (Newman and Thornley, 1996). Currently, the municipal plan is divided into two parts: a four-year action plan determined in part by the annual budget cycle, and a twelve-year physical plan reviewed every four years in relation to the electoral cycle (Holt-Jensen, 1994). Municipalities also have to produce a local or regulation plan (*reguleringsplan*), which is binding on individuals and specifies the use of each plot and the layout and design of the development (Newman and Thornley, 1996).

Regional policy and regional incentives

For generations, the three most northerly counties of Norway (Nordland, Troms and Finnmark), have possessed one of the most unfavourable environments for economic development in Europe. A very low population density and an inaccessible settlement structure within a remote region imposed a substantial constraint on economic growth, compounded (until the 1970s) by high levels of net out-migration to southern regions of greater opportunity (Williams, 1987). The North Norway Plan of 1952 promoted, through the use of investment subsidies, loans and tax privileges, the development of a year-round fishing industry by modernising shipping fleets and constructing new processing plants, and also encouraged the occupational mobility of labour from largely peasant agriculture to modern manufacturing (Williams, 1987). In the 1960s, with the establishment of the Regional Development Fund (RDF), regional policy focused instead on infrastructure investment, the provision of transport subsidies and the development of industrial estates. Despite the rapid development of oil exploration and refining since the 1970s, areas in the south, such as the county of Rogoland with Stavanger, have been the principle beneficiaries, and, in general regional policy, failed to enhance the development potential of the north (Williams, 1987). Out-migration from the north to the major urban centres of the south, having temporarily subsided in the 1970s, thus recurred in the 1980s.

The RDF, as the principal funding agency for regional development until 1993, allocated the financial means for development through the budget of the Ministry of Local Government and Labour to the appropriate counties and municipalities within designated development zones. In Finnmark and North-Troms, the upper level of investment grants in the early 1990s ranged from 40 per cent in Finnmark and North-Troms to only 15 per cent in western Norway, whereas in eastern and southern Norway and Rogaland, loans were the only form of assistance. In total, the development zones contained 32 per cent of the national population (Sýkora, 1994a). With the Norwegian Industrial and Regional Development Fund succeeding the RDF in 1993, the responsibility of stimulating economic activity and employment in the peripheral regions will be accompanied by the function of fostering internationally competitive enterprises.

On a per capita basis, regional incentives are comparable with those available in the most disadvantaged regions of the EU, although the degree of economic difficulty is less severe in Norway than in, for example, southern Italy or Greece. Had Norway joined the EU in 1995, along with Sweden and Finland, she would undoubtedly have had to adjust to EU priorities in which support for rural areas with a low population and emigration is normally considered less important than incentives for industrial restructuring in regions with high unemployment and low income per capita. Nevertheless, Norway would have been a recipient of Objective 6 funding aimed at promoting the development and structural adjustment of regions with extremely low population densities. An Objective 6 allocation of 384 million ECU (at 1994 prices) or 125 ECU per capita could have been widely distributed in an area containing some 587, 000 people or only 13.8 per cent of the total population (CEC, 1994c).

Finland

Influenced much by the Swedish legal and administrative system (Finland was part of Sweden in the seventeenth and eighteenth centuries), Finland, like Sweden, introduced a highly centralised welfare state after the Second World War, with local authorities often acting merely as agents of central government (Sotarauta, 1994). In planning, however, lower-tier administrations had comparatively more power. Although legislation of 1958 established a hierarchy of plans, there was a complete absence of a national plan, and while regional plans were intended to provide guidance for municipal general plans (which, in turn, provided the framework for detailed plan), a two-way dialogue emerged. Thus, although the upper tiers gave directions and recommendations to the municipality, planning initiatives and information emanated from the municipality (Newman and Thornley, 1996).

While central government planning policy is implemented by the eleven counties in a supervisory capacity, regional plans (reflecting central government aims and objectives) are prepared by nineteen Regional Planning Associations elected in proportion to the political distribution of constituent municipal councils (Mansikka and Rautsi, 1992). Regional plans for a 4 to 5-year period, indicate the broad distribution of land uses and direction of communication development, and are submitted to central government for approval. Thereafter, although regional plans are not legally binding, it is expected that municipal plans will be largely complementary (Newman and Thornley, 1996).

As elsewhere in Scandinavia, lower-tier authorities have a dominant position in the planning system and approve their own plans, albeit within frameworks provided by the upper and middle tiers of government. Since 1992, 461 municipalities have been divided into 88 joint intermunicipal boards to undertake spatial planning at the local level, thereby diminishing the role of the counties as more of their functions are transferred to the new lower-tier boards (European Commission, 1994). In addition, the boards have the responsibility of producing structure plans (*Yleiskaava*) for the whole of their areas which, together with associated regulations, are legally binding on lower-tier government. Structure plans are essentially concerned with broad land-use zoning and, in the past, were ratified by central government (Newman and Thornley, 1996). Although this control is diminishing, some lower-tier authorities, in their desire to protect certain uses from development (for example recreation and conservation areas), might seek ratification of parts of their plans which, if successful, would make them legally binding on individuals (Mansikka and Rautsi, 1992).

When development is about to take place, a detailed town plan (an *Asemakaava*) is required and is legally binding on individuals. If development complies with the appropriate land-use zone indicated by the plan, planning permission is deemed to be provisionally granted but a building permit is required before development can be finally authorised.

In response to grassroots demands for greater public participation in the planning process, the Planning and Building Act of 1990 aimed both to increase participation and to encourage sustainable development. Under the

Act, municipalities are thus required to prepare annual planning reports specifying developments in the course of preparation (together with those that have been completed), and the Act also introduced a procedure whereby municipalities, whilst being obliged to seek ratification of their general plan, could subsequently prepare detailed plans without having to submit them for approval (Virtanen, 1994; Newman and Thornley, 1996). Clearly, the effects of the legislation were to decentralise planning power from the upper-tier authorities to the municipality, and to liberalise and hence hasten the pace of the processes of planning and development (Haila, 1990). While a further review of the planning system in 1993 focused on the environment and the needs of disadvantaged segments of the population, it also proposed that decentralisation should be accelerated by central government only examining rather than ratifying regional plans (Virtanan, 1994; Newman and Thornley, 1996).

Regional policy and regional incentives

In much of Finland, although the standard of transport and communication infrastructure is good, the forest and lake environment is particularly sensitive to air and water pollution emanating from mining, agriculture, logging and pulp processing. Above all, large areas of the country are very distant from the major urban areas, and population density is as low as four persons per square hectare. In these remote areas, unemployment in the 1990s often exceeded 24 per cent compared with 20 per cent nationally, in part reflecting the narrow employment base.

To compensate for difficult natural living and working conditions, low population densities and remote locations, Objective 6 funding, amounting to 450 million ECU, 1996–99, was made available by the Commission to enhance the position of Finland in the world economy, widen job opportunities, support existing agriculture, and protect the natural environment. Funding was available to large parts of northern and eastern Finland, comprising mainly the counties of Lappi and Oulu (see Figure 5.3), and was to be particularly targeted at the development of businesses and company competitiveness, at the development of human resources and at agriculture, natural resources and the environment (Williams, 1996). Thus, in contrast to the non-Nordic countries of the EU, a principal aim of regional policy in Finland, as in Sweden, has been to maintain the level of population, employment and incomes in the remoter areas of the country, rather than to focus on disparities in output and employment in the main centres of population. In the less remote areas of Finland, Objective 2 areas were eligible for grants of up to 179 million ECU, 1996–99, to facilitate the economic conversion of areas of industrial decline, and 135 million ECU were available to Objective 5b areas over the same period to facilitate rural development. Applicant counties were obliged to submit development and conversion plans to the central government for consideration as a preliminary

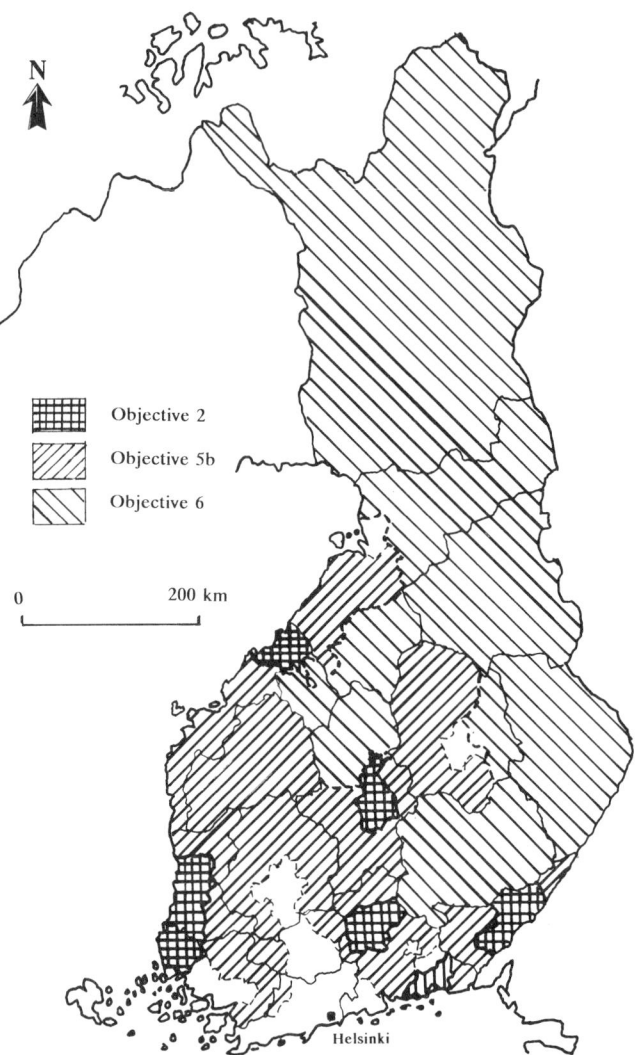

N

Objective 2

Objective 5b

Objective 6

0   200 km

Helsinki

Figure 5.3 Designated Structural Fund areas, Finland, 1995–99

to seeking funding from the Commission. In total, 53.6 per cent of the population of Finland lived in areas benefiting from structural funding: 15.5 per cent in Objective 2 areas, 21. 5 per cent in Objective 5b areas and 16.6 per cent in Objective 6 areas (CEC, 1996b).

6 Unitary states with planning powers devolving to the regions

A number of unitary states in the European Union have undergone, or are undergoing, a process of reform to establish an elected layer of regional authorities between the central state and local government. Among these countries, Portugal remains the most centralised, despite its constitution recently granting autonomy to the islands of the Açores and Madeira. The five regions on the mainland – governed by branches of the central state – are, nevertheless, appropriate in scale for the creation of an intermediate tier of elected government in due course. Until very recently, it could have been argued that the United Kingdom was also one of the most centralised unitary states in Western Europe, but with a Welsh Assembly, a Scottish Parliament and an Assembly in Northern Ireland being in place by the turn of the century, the United Kingdom has become a 'devolving' rather than a 'classic' unitary state. With the possibility of elected assemblies established in the English regions in the future, the devolutionary process will be taken a stage further. With its twelve provinces, the Netherlands has had an intermediate level of decentralised government for some time, yet to an extent the Dutch state is still in a process of devolution. It was decided in the early 1990s to establish a small number of metropolitan regions, with elected governments, to replace a number of local and provincial authorities, but by the late 1990s these had not been set up – opening a window of opportunity for the provinces to enhance their own powers of government. Through creating a new layer of twenty-two regions since the early 1980s, France has the most far-reaching system of elected decentralised government of any unitary Western European state (Bullman, 1997). It is doubtful, however, whether there will be further opportunities for decentralisation. Regional authorities in France neither have any legislative power nor any power over local government in terms of its structure, supervision or finance, but, by stimulating and coordinating the important players in a region, can create the conditions for regional development – specifically with the fields of regional economic planning and economic aid, spatial planning and education at secondary level and above. This chapter will examine the planning systems of the above countries, commencing with a review of government and planning in the most centralised devolving state – Portugal, and concluding with an exploration of the complexities of decentralised government and planning in France.

Portugal

Under the highly centralised Salazar administration (1927–74), a formal system of physical planning was gradually set in place, but it was one which focused on urban areas (ignoring the regions) and lamentably failed to control illegal development during the latter years of the regime. Following a change to democratic government in the mid-1970s, new legislation and a hierarchy of plans emerged.

Central government is currently responsible for passing legislation, producing national plans for socio-economic development and co-ordinating administration at a regional level. With regard to physical planning, it is responsible for producing a plan for the national road network with the aim of stimulating local development, it conversely restricts development in the most productive agricultural areas and in environmentally sensitive parts of the country especially on the coast, and it approves municipal plans, or, where these have yet to be approved, it acts as the development control authority (Newman and Thornley, 1996).

Until recently, central government exercised its power across the country through the medium of eighteen district offices, each under a governor in conventional Napoleonic style. Although Portugal is still very largely a unitary state, recent reforms have replaced the districts with seven regions – two of which, the Açores and Madeira are autonomous, with special rights, elected government and limited legislative power (Wiehler and Stumm, 1995). The new regions of mainland Portugal are, however, under the administration of Regional Co-ordination Commissions (RCCs), branches of central government established to co-ordinate and implement policy at regional level. RRCs have a responsibility for producing regional physical plans (*Planos Regionais de Ordenamento do Território*) with the aim of providing a spatial framework for both investment and land-use planning at municipal level. The RRCs also have an enabling role in providing technical help to municipalities if it is considered necessary to integrate planning on a sub-regional scale (Newman and Thornley, 1996).

At the municipal level, the power of the unitary state is still very evident. Although each of the 305 municipalities has the responsibility for producing a municipal plan (a *Plano Director Municipal*), the plan (concerned with physical and socio-economic development across the whole of the municipal area) has to be submitted initially to the Technical Commission – composed of representatives from the Regional Commission and other state bodies. It subsequently becomes the object of a public inquiry, and only then is it considered for approval by both the Municipal Assembly and the Minister of Planning and Regional Development (Newman and Thornley, 1996).

Although the approval of a plan is often linked to central government funding of municipal activity and to conferred powers of expropriation, only thirty-three municipal plans had been produced by 1993 (EC, 1994b). This, arguably, could be attributable to the onerous approval process which slows down the preparation of plans until they have been produced to a 'satisfactory' standard, but might also be due to the lack of planning expertise at municipal level and the unwillingness of municipalities to produce plans that are sufficiently regulatory and inflexible to secure government approval. Where municipal plans have been approved, the municipalities are empowered to produce more detailed urban plans for parts of their area (*Planos de Urbanizão*), and detailed layout plans (*Planos de Pormenor*) in collaboration with the 3850 parishes (*freguesias*), while some of the better-resourced urban authorities are beginning to produce strategic plans to provide a context for the implementation of their municipal plans (Newman and Thornley, 1996).

Although developers are obliged to comply with the requirements of these plans, there was a need to legitimise development which had contravened planning law in the past. With the assistance of *comissariados*, established by central government, plans were prepared in the 1970s to legalise unauthorised development retrospectively, and technical help had been made available to the municipalities for this purpose. Except in Lisbon, the process of legalisation has, at best, been patchy (Williams, 1984). Illegal development must be seen in the context of a slow and centralised system of planning, and where, at a local level, plans are too regulatory and inflexible (Vasconcelos and Reis, 1994).

Regional policy and regional incentives

With a per capita income only slightly above that of Greece, Portugal is the second poorest country in the EU. While the Açores and Madeira are disadvantaged by population decline and a deficiency of productive economic activity, in mainland Portugal there are marked spatial variations in living standards, particularly between the relatively prosperous Atlantic littoral and the impoverished agricultural areas of the interior. However, even the richest region, Lisboa e vale do Tejo had a per capita GDP of only 77 per cent of the EU average in the early 1990s, whilst the GDP per capita of the poorest mainland region, Alentejo, was only 34 per cent of the average. In addition to regional disparities, the whole country – to a greater or lesser extent – suffers from a remote location in Europe, inadequate infrastructure, an underdeveloped urban network, a weak industrial base, a small and non-specialised national market, low levels of education and vocational skill, and an inefficient agricultural sector (Williams, 1996).

Following Portugal's entry into the European Community in 1986, and in response to the need to bring the country more into line with other members of the EC, a new system of regional incentives was immediately introduced comprising a capital grant, an employment premium and support for innovation – providing assistance of up to 33 per cent of eligible expenditure. In 1988, regional aid was substantially increased following discussions with the European Commission – with up to 65 per cent of investment becoming eligible for grant assistance, and, in addition, support was provided to enhance the business environment and to encourage the development of tourism. Further amendments were made to the grant system in 1989 targeting aid to the worst-off areas (where 75 per cent grants were available), while a new system of grant aid and interest free loans – SINDEPEDIP (*Sistema de Incentivos a Estrategias de Empresas Industrias*) – was introduced to facilitate 'softer' or intangible asset investment elsewhere in the country. In 1994, Law 193/94 introduced a new regional incentive scheme – SIR (*Sistema de Incentivos Regionais*) – to help fund the creation and modernisation of small and indigenous firms in the underdeveloped areas of the interior, while larger projects continue to be assisted by SINDEPEDIP aid which takes the form of a grant of up to 70 per cent of eligible investment and an interest free loan covering 80 per cent of all other eligible expenditure (see Yuill, *et al.*, 1996). Separate forms of fiscal and financial aid are available to encourage investment in the autonomous but economically disadvantaged regions of the Açores and Madeira.

Since joining the EC, Portugal has been a major recipient of Structural Funds. For the period 1994–99, Portugal received Objective 1 funding of 14 billion ECU – the whole country (including the autonomous regions) being eligible for this form of aid (CEC, 1996b). Clearly, there was a need for a substantial amount of public sector investment on a scale beyond the resources of the national authorities (Williams, 1996), but to ensure that investment was targeted effectively, the Community Support Framework set four priorities for Objective 1 funding:

1   An improvement in economic competitiveness through the development of the infrastructure, particularly transport, energy and telecommunication networks, and the modernisation of manufacturing, retailing, agriculture, fisheries and tourism (59 per cent of the budget).
2   The strengthening of the regional economic base through support for local authorities, and programmes of urban and rural development (18.5 per cent of the budget).
3   The development of education and training, and the promotion of employment (15.5 per cent of the budget).
4   Environmental action, urban renewal and improvements to health and social provision to improve the quality of life and social cohesion (7 per cent of the budget).

Within the context of these priorities, sixteen Operational Programmes were introduced – one for each of the seven regions, six for various sectors such as infrastructure and industrial development, and the remainder for other matters such as municipal development projects.

United Kingdom

Developed over the course of the twentieth century, the planning system of the United Kingdom has tended to be highly centralised, with overall responsibility in the last decades of the millennium lying with the Secretaries of State for the Environment, and for Scotland, Wales and Northern Ireland. However, at the time of writing, a Labour government, with a secure majority of seats in the House of Commons, is committed to creating nominated regional chambers in the English regions as a prelude to establishing directly elected assemblies if there is clear public support for doing so. A Welsh Assembly and a Scottish Parliament, moreover, are scheduled to be in place by the year 2001, and in Northern Ireland constitutional reform is an outcome of the 1993–98 peace process. With devolution, not only would the responsibility for town and country planning be transferred to accountable government in Wales, Scotland, Northern Ireland and possibly the English regions, but the new administrations will have the task of integrating physical and economic planning on a regional, national or provincial scale. Since there are notable differences between the planning system in the different parts of the United Kingdom and because the pace of devolution will vary geographically, each of the constituent parts of the United Kingdom will be considered in turn.

England

Centralised power

After several decades of evolution, the English planning system remains highly centralised but, in contrast with many other countries, there is a lack of a plan at national level. Although at lower tiers of government, county councils, district councils, the metropolitan and London boroughs and the unitary councils all have various planning functions, overall responsibility lies unequivocally with the Secretary of State for the Environment. First and foremost this involves competence for new legislation governing the planning system. Although planning legislation in England has a long history dating back to the Housing, Town Planning, etc. Act of 1909, the Housing and Town Planning Act of 1919 and the Town and Country Planning Act of 1932 (each focusing on development control by zoning), it was only after the Second World War that comprehensive planning legislation was introduced – the Labour government's Town and Country Planning Act of 1947. Under the Ministry of Town and Country Planning, the 1947 Act established the basic principles of the national planning system: development plans, development control and central government supervision – principles which were enshrined in subsequent legislation, most notably the Town and Country Planning Act of 1968 and the Planning and Compensation Act of 1991.

Centralised power is further exercised when the Secretary of State calls in development plan proposals for approval when intervention is considered necessary, determines planning applications that he has 'called in' for his own decision, and employs the Planning Inspectorate to determine planning appeals, deal with enforcement appeals and handle public inquiries into plan proposals drawn up by local authorities.

Since 1988, central government has, in addition, provided guidance on planning matters by issuing a series of notes on planning policy guidance (PPGs), mineral planning guidance (MPGs) and derelict land grand advice (DLGAs) and from 1989 produced a separate series of notes on regional planning guidance (RPGs). Guidance notes set out the broad guidelines councils should take into account when determining planning applications and which the Secretary of State and the Planning Inspectorate should consider when determining appeals.

However, in the 1980s, central government in pursuit of 'Neo-Liberal' objectives made 'a strong side swipe at mainstream traditional town and country planning' (Cullingworth and Nadin, 1994: 67). Enterprise zones, urban development corporations and simplified planning zones were introduced in an attempt to regenerate the inner cities and other areas in need of re-development, but local authority powers to exercise development control were reduced or withdrawn. Under the Local Government, Planning and Land Act of 1980, the Secretary of State was empowered to designate enterprise zones (EZs) in which business activity would be promoted by means of exemption from rates on industrial and commercial property, a 100 per cent tax allowance for capital expenditure, fewer demands for information, and a general permission to develop. The 1980 Act also empowered the Secretary of State to establish urban development corporations (UDCs) to usurp the development control functions of local authorities. UDCs were run by unelected Boards of Directors appointed by the Secretary of State and answerable only to him in the conviction that more development would be attracted to the designated areas if local residents and their political representatives were excluded from the planning process. In contrast, the Housing and Planning Act of 1986 empowered local authorities to declare simplified planning zones (SPZs) where planning permission was deemed to be granted for development or classes of development specified in the scheme. Although EZs, UDCs and SPZs were to have an operational life of approximately ten years, in a large number of locations they represented a serious attempt by the government to reduce the traditional planning responsibilities of local authorities through a reversion to market and an intensification of central power.

The regional dimension

The regional tier of government is as yet absent in England, although throughout much of the twentieth century there have been numerous attempts to introduce various forms of governance and planning on a regional scale. The need for a rational approach to administration was strongly argued by the Fabian Society (1904), Patrick Geddes in *Cities in Evolution* (1915), C.B. Fawcett in *Provinces of England* (1916) and G.D.H. Cole in *The Future of Local Government* (1924), but a regional tier of government lacked political support in Westminster. Instead, throughout most of the twentieth century, the emphasis was on sub-regional or city-regional planning undertaken within the context of local government. Sub-regionally there have been four main surges of planning of which the first – the 'experimental era' – commenced after the First World War and continued until the beginning of the Second (Wannop, 1995). During this period there were twenty regional surveys and reports undertaken in, for example, East Kent in 1925, Leeds and Bradford in 1926, Hertfordshire in 1927, South Buckinghamshire and Thamesside in 1928, Greater London and North-East Lancashire in 1929, the North Riding and Cambridgeshire in 1934 and Bedfordshire in 1937. More importantly, as many as thirty-seven regional planning schemes were prepared during the inter-war years in, for example, Doncaster in 1923, West Middlesex in 1924, Manchester and District in 1926, Lancaster and Morecambe in 1927, Mid-Surrey in 1928, West Sussex in 1929, Bath and Bristol 1930, Oxfordshire 1931, Leicestershire 1932, Greater London 1933, East Suffolk in 1935 and Harrogate and District in 1937 (see Wannop, 1995: 22–24). It must not be assumed, however, that during the 1920s to 1930s regional planning was strategic or visionary. Massey (1989) explains that its rationale reflected the wish of local authorities to implement town planning legislation within a convenient regional or sub-regional context. It can also be suggested that 'regional planning . . . was taken up by local authorities because its measures seemed not to hurt anyone, they seemed full of promise, they cost little and offered co-operation rather than conflict' (Wannop and Cherry, 1994).

Prior to the Second World War, regional governance became important as a matter of security. A comprehensive regional organisation was set up under the command of Regional Civil Defence Commissioners in 1939 to co-ordinate government departments in nine regions of England in the event of a collapse of administration from London (Wannop, 1995). After the war, regional governance was central to the process of reconstruction. In 1946, the Treasury defined a set of nine standard regions, in which central government departments were expected to operate in order to facilitate co-operation between regional officials (Mackintosh, 1968). Between 1946 and 1951, the standard regions also formed the basis for the spatial organisation of the Hospital Boards, Railway Boards, Area Gas and Electricity Boards (all set up as corporate entities by Act of Parliament), and regional bodies such as the BBC, the Coal Board and the Central Electricity Board (each independently organised on a regional basis).

The regional structures so created were far from integrated and failed to further the development of economic planning. There was no simultaneous development of regional physical planning and there were few calls for elected regional authorities. Instead, there was a second surge of sub-regional planning focusing on the major conurbations. With its emphasis on regional imbalance, the *Report of the Royal Commission on the Distribution of the Industrial Population 1940* (the Barlow Report) called for a national planning authority not only to balance the inter-regional distribution of industry but also, intra-regionally, to 'decentralise people and industry from the congested and unfit metropolitan areas' (Wannop, 1995: 7). Although a 'national planning authority' was set up in the form of the Ministry of Town and Country Planning in 1943, its terms of reference were confined to land-use planning, with the Board of Trade assuming responsibility for industrial location in 1945 – a bifurcation which (in a modified form) bedevilled regional planning in England throughout the rest of the twentieth century. The Barlow Report, nevertheless, inspired the production of two regional plans of considerable significance during the Second World War – Sir Patrick Abercrombie's County of London Plan of 1943 and the Greater London Plan of 1944. Although these paved the way for the new towns programme and other overspill measures in the South East during the late 1940s to early 1950s, other regional plans prepared for Hull in 1943, Merseyside in 1944, South Lancashire and North Cheshire in 1947, the West Midland Conurbation in 1948 and the North East in 1949 were to a varying extent 'cast in the mould of the regional surveys characteristic of earlier times' (Wannop, 1985: 8). Partly because these plans were inadequately prescriptive, and partly because county councils and county boroughs became pre-occupied in preparing development plans for their areas under the Town and Country Planning Act of 1947, regional planning virtually disappeared from the agenda for well over a decade – an absence compatible with the Churchill government's antagonism to planning at any spatial level.

However, with a marked increase in unemployment in the early 1960s, most notably in the North, and with current projections of future population growth far exceeding the capacity of local authority development plans particularly in the South-East, the need for planning on a regional scale was again accepted and a third surge in planning activity commenced (Wannop, 1995). With regard to alleviating high unemployment, the consequential *The North East: a programme for regional development and growth* Ministry of Housing and Local Government, 1963) therefore set out a number of positive planning programmes – both economic and physical – for regional regeneration whilst, to contain population growth, the *South East Study* (Ministry of Housing and Local Government, 1964) prepared the ground for the development of a series of large new towns in the region in the late 1960s and 1970s.

In the early 1960s, a further cause for concern within the arena of regional planning was the absence of an elected strategic authority for the London conurbation – an area of about 620 square miles with a population of nearly eight million. Consequently, under the London Government Act, 1963, the Greater London Council (the GLC) came into operation in 1965 with tax-raising powers and with responsibility for overall planning, main highways, traffic control, overspill housing and ambulance and fire services. The GLC was given the task of producing a strategic plan (the Greater London Development Plan) for the whole of the Greater London area, and within this framework the thirty-two new London boroughs were required to prepare local plans in addition to the provision of a wide range of other local services.

A year after the return of a Labour administration in 1964, regional planning appeared to be given yet another fillip by the setting up of Regional Economic Planning Councils (REPCs) and Regional Economic Planning Boards (REPBs) in each of the six newly defined English planning regions (see Figure 6.1). Under the overall control of the newly created Department of Economic Affairs, REPCs comprised representatives from business, the trade unions, local government and the universities – each appointed by the Secretary of State, while the REPBs consisted of regionally based civil servants under the chairmanship of a Department of Economic Affairs official. Although REPBs were initially given the task of preparing regional plans in liaison with the REPCs, within months this responsibility was transferred to the REPCs (partly to enable the government not to commit itself to implementation). Henceforth the councils, in addition to preparing regional plans, advised the government on the measure necessary to implement regional plans on the basis of information and assessments provided by the REPBs, and offered advice on the regional implication of national economic policies. However, in the early 1970s, following the abolition of the Department of Economic Affairs, REPBs were disbanded, and the functions of the REPCs were downgraded, although they continued to produce plans, often in association with the regional offices of government departments and local authorities. Thus, either alone or in collaboration, the REPCs between 1965 and 1979 produced full regional studies and strategies for all the English regions (except for the West Midlands, where local authorities took the initiative), and in addition produced sub-regional studies and plans for a number of pressured non-metropolitan city regions such as Coventry, Solihull, Warwickshire and South Hampshire. However, the Department of the Environment (responsible for the REPCs after 1970) and the Department of Trade and Industry were unable to integrate land-use planning and economic development – rendering planning at the regional level a largely physical exercise, notwithstanding attempts to incorporate tables, figures, distribution maps and economic analyses into most regional plans by the 1970s (Wannop, 1995).

In the 1970s, the reform of local government could have had a major

Figure 6.1 Economic planning regions, Great Britain, 1965

impact on regional planning. There had been calls to establish a regional tier of government in the 1960s, particularly after the establishment of economic planning regions and the setting up of the REPCs and REPBs. Mackintosh (1968), for example, proposed the creation of nine elected regional councils in England with responsibilities for regional planning, highways, large-scale housing development and a host of other activities which could more appropriately be administered on a regional rather than local scale, and subsequently the Redcliffe Maud Report (Royal Commission, 1969a) recommended that sixty-one English counties be grouped into eight provinces (largely coinciding with the economic regions) in each of which an assembly of county representatives would have responsibilities for regional planning and other strategic matters – the implication being that they would subsume the role of the REPCs. However, under the Local Government Act of 1972, the Conservative administration failed to take the opportunity of creating a regional tier of government and instead made comparatively minor modifications to the existing system of local government. The Conservatives also ignored the recommendations of the subsequent Kilbrandon Report (Royal Commission, 1973) that eight non-executive regional councils – comprising local government councillors – should be set up in England to replace the REPCs and to provide interfaces between central and local government, and local and regional interests (Wannop, 1995). In contrast, the planning of strategic public services on a regional scale was advanced by the creation of fourteen Regional Health Authorities and nine Regional Water Authorities in 1974 with the aim of facilitating a more rational allocation and a more efficient distribution of services.

Following a comparatively short period of Labour administration, 1974–79, in which further strategic regional plans and sub-regional studies were published, a Conservative government was returned to office under the premiership of Mrs Thatcher. Wedded to the market mechanism, it had 'an antipathy to planning unmatched since the Churchill government of 1951' (Wannop, 1995: 19), and immediately abolished the REPCs and terminated the production of regional plans. However, with growing concern in the 1980s about intra-regional disparities, green issues, loosely-controlled housing development in the countryside of the South-East, increasing traffic congestion in Greater London and beyond and the impact of the Channel Tunnel and the European Community Single Market, regional planning inevitably became a political necessity. Although standing conferences of local planning authorities for the South-East and West Midlands date from the 1960s and were relatively successful in co-ordinating planning at a regional level, in other regions there was comparatively little sustained collaboration. However, by the mid-1980s, there was a new and urgent willingness to collaborate in most regions – an activity welcomed by the government in its Green Paper, *The future development plans*, (Department of the Environment Welsh Office, 1986), and by 1992 – with the formation of a standing conference in the North West – conferences covered the whole of England.

The 1986 paper signalled the introduction of stronger regional guidance 'as an integral element of the statutory planning system' (Tewdwr-Jones, 1996: 32) with regional planning, henceforth, being no longer advisory, but statutory. In the subsequent White Paper, *The Future of Development Plans* (HM Government, 1989), it was clear that the government intended to realise its regional planning objectives by 'providing regional planning guidance where necessary to assist in the preparation of new statements of County Planning Policies and District Development Plans' (Tewdwr-Jones, 1996: 32). PPGs 9, 10 and 11 had already been issued for the South East, the West Midlands and Merseyside in 1988 for this purpose, and over the following six years. RPGs (concerned more with the strategic and visionary aspects of the plan process) were issued for Tyne and Wear, West Yorkshire, London, Greater Manchester, South Yorkshire, East Anglia, the Northern Region, the East Midlands, the South East and the South West. However, the Department of the Environment was not anxious 'to see the embryonic regional planning system develop into a more effective form of regional planning that took account of other non-land use matters' (Tewdwr-Jones, 1996: 33–34), let alone support the creation of the large regional planning authorities.

However, by the early 1990s, a more and more centralised system of government ironically created the need for a reinforced regional level of public administration. The Major government consequently established nine 'integrated regional offices' (IROs) in England to co-ordinate the activities of the Department of the Environment, the Department of Transport, the Department of Trade and Industry and the Department of Education and Employment (see Figure 6.2). The principal function of each office

Figure 6.2 New regional offices, England, 1994

was to handle bids for the Single Regeneration Budget (worth £240 million in 1995–96), bids for industrial aid and bids for aid from the EU Structural Fund – prior to being submitted to Whitehall. But IROs were unelected, and had more in common with the French system of prefectures than with devolved representative government.

Yet, for many years, it has been posited that regional economic and spatial planning necessitates – on the grounds of efficiency and account-ability – the creation of an elected regional tier of government. As Luttrell (1987) pointed out, Britain has become 'the apotheosis of a unitary state', but since this condition has arguably weakened the regions (both north and south) and is of dubious benefit to the national economy, it is necessary to establish, as in other countries, an 'institutional framework (that) . . . enables the regions to help themselves, to have the powers and resources to do so, and to become decision-making centres in their own right' (Luttrell, 1987). Luttrell (and in broadly similar terms, the Town and Country Planning Associations, 1989) proposed that, first, directly elected councils should be set up in each of the eight regions of England, and that they would receive powers devolved to them from central government. County councils would become redundant since their responsibilities would be absorbed by lower-tier authorities or the new regional councils. Second, a newly-created ministry, or an existing ministry, would need to be charged with the responsibility of co-ordinating intra-regional policy in each English region in full collaboration with the regional council, while regional councils would implement their regional strategies partly through the medium of newly established development agencies and partnerships with the private sector. Third, using central government grants and precepts on local authorities, regional councils would need to finance both strategic and physical planning, social housing, highways, public transport and other key public services and, ultimately, regional councils would have tax-raising powers of their own.

The Labour Party also favoured a considerable degree of devolution to the regions. In the late 1980s, it proposed setting up ten elected regional assemblies (each containing about 5 million people) to exercise power devolved from central government, and to 'absorb under democratic control, the functions exercised by non-elected boards and quangos' (Labour Party, 1989: 57) – proposals which were incorporated in the party's 1992 election manifesto. At the 1997 General Election, Labour – more cautiously – pledged that it would initially establish regional chambers to co-ordinate economic development, planning, transport, bids for European funding and land-use planning. Since the demand for directly elected regional assemblies varied across the country, Labour thought it wrong to impose a uniform system of devolved government, but in time it would introduce legislation 'to allow people, region by region, to decide in a referendum whether they want directly elected regional government' (Labour Party, 1997). If, as a result of popular consent, regional assemblies were to be established in

England, Labour thereby would not be adding a new tier of administration to the existing hierarchy of government since county councils would be abolished and district councils would be grouped into unitary authorities responsible for a wide range of local government services.

As a possible first step towards implementing these pledges, Labour's Regional Development Agencies Bill, 1997, proposed establishing agencies in the regions of England to promote inward investment, help small businesses and co-ordinate regional economic development. The agencies would be expected to work in liaison with the IROs, training and enterprise councils and existing non-statutory regional chambers established by local authorities. Only time will tell whether the relationship between the agencies and the IROs will work effectively and whether devolved government will eventually emerge from this initiative.

Planning and local government

The system of land-use planning in England which was to prevail throughout the remainder of the twentieth century was set in place by the Local Government Act of 1972. Under this legislation, Greater London and six metropolitan counties (the West Midlands, Merseyside, South East Lancashire, South Yorkshire, West Yorkshire and Tyne and Wear) together with thirty-nine 'shire' counties, sixty-six London boroughs and metropolitan districts and 264 district councils became responsible for the preparation of development plans at a local level and for exercising development control. The Town and Country Planning Act of 1971 (in anticipation of local government reform) introduced new development plans to be prepared and implemented at local authority level. County councils were obliged to produce structure plans containing strategic policies on a broad range of economic, physical and social issues across the county, whereas district councils and the London boroughs were encouraged to prepare local plans containing specific land-use allocations to facilitate both detailed forward planning and development control (Tewdwr-Jones, 1996). A local plan could cover the whole or part of a district (or London borough) or take the form of an action area plan or a subject plan.

During the second term of Mrs Thatcher's premiership, the planning system was constantly in a state of flux and uncertainty as the Conservative administration attempted to 'roll back the frontiers of the state', to allow the market to increasingly dominate the economy. The GLC and the six metropolitan counties were abolished along with their strategic planning role under the Local Government Act of 1985, and their planning responsibilities transferred to the London boroughs and the metropolitan districts. Within these authorities, but at the expense of strategic planning, unitary development plans were introduced to subsume the functions of the former structure and local plans. There was also a reduction in development control since, in consequence of the White Paper, *Lifting the Burden* (HM Government, 1985), material considerations (such as the need to create employment) rather than development plans formed the basis for judging an application for planning permission – a *volte face* which heralded a spate of appeals by developers against the decisions of local planning authorities based on the objectives of their development plans. Both the Green Paper, *The future of development plans* (Department of the Environment/Welsh Office, 1986) and the White Paper of the same title (Department of the Environment, 1989), moreover, proposed abolishing structure plans – proposals which were subsequently rejected by the Secretary of State in the autumn of 1989.

Under the Major government the direction of policy seemed to change and planners were encouraged 'to enthuse about the start of a new era' (Newman and Thornley, 1996: 120). The Planning and Compensation Act of 1991 aimed to speed up the preparation of development plans, eliminated the need for the Secretary of State to approve structure plans, required district authorities to mandatorily prepare district-wide local plans for the whole of their areas, and increased the importance of development plans in decision-making processes (Tewdwr-Jones, 1996). Unlike policy in the late 1980s, the Act 'stated that planning decisions should be taken "in accordance with the plan unless other material considerations indicated otherwise"' (Newman and Thornley, 1997: 120). At the same time, the Local Government Commission – appointed under the Local Government Act of 1992 – began a review of English local government in an attempt to arrive at a structure which most appropriately reflects community loyalties. By March 1995, following an unfinished review of twenty-nine counties, the government decided to abolish five (replacing them with a total of nineteen unitary authorities), introduce a 'hybrid' system in fifteen other counties (involving a mixture of unitary authorities and the remains of the two-tier system of local government), and retain the two-tier system in the remaining eighteen counties.

Yet, underlying all these changes, centralised power was being maintained or increased. Central government, aware of its ability to determine local policy through its Planning Policy Guidance, continued to issue new and revised PPGs and, in so doing, increasingly stamped its mark on structure, local and unitary plans, while at the same time central government continued to call in development plans or use the appeals system to ensure that local authorities only exercised their planning responsibilities with the explicit consent of the state.

In the final years of the twentieth century, however, there was a reaction against the centralised power of the state. In keeping with its 1997 General Election pledge, the Labour Party, following its electoral success, not only conducted referendums on devolved government in Wales and Scotland (see pp. 102 and 106) but also held a referendum to determine whether or not an elected regional authority with its own mayor, should be set up for the capital – London (since the abolition of the GLC in 1986) being the only western centre of government without its own city-wide administration. In May 1998, 73 per cent of the poll registered a vote in favour of the formation of a Greater London Authority, the government consequently proceeding to legislate to establish a regional tier of government for the capital with strategic planning powers and to create the office of an elected mayor (an innovation for the metropolis) – the first elections taking place in May 2000.

Wales

Since its union with England under the statutes of 1536 and 1542, the principality has shared with England a broadly common system of central and local government. In the 1880s, however, there was a call for a Welsh Parliament from Cymru Fydd (Wales of the Future) and in the following decade there were proposals from the Liberal Party to superimpose a top tier on the county council system to deal with matters of common interest to Wales, to create a devolved executive in the principality and to appoint a Secretary of State (Mackintosh, 1968). With the industrialisation of South Wales, further integration with the English economy, and the collapse of the Liberal Party, demands for home rule diminished; and, in the inter-war period, record unemployment in the south led to a preference, among much of the population, for a Labour victory in the general elections of the 1920s to 1930s than for nationalism. Central government, nevertheless, was already introducing policies which eventually put devolution on the political agenda (Mackintosh, 1968). As early as 1907, a separate Welsh Department was formed within the Ministry of Education, and by 1945 most major departments of state contained a Welsh office and controller. A Conservative government in 1951 made the Home Secretary also Minister for Wales, and in response to a recommendation of the Advisory Council of Wales in 1957, a Labour government in 1964 established the Welsh Office under a Secretary of State with a seat in the Cabinet (Mackintosh, 1968).

In recent years, more and more responsibilities over a wide field have been decentralised from Whitehall to the Welsh Office in Cardiff – health, community care, education, agriculture, forestry and fisheries, industrial development, urban renewal, local government, housing, water and sewerage, environmental protection, nature conservation and the countryside, and town and country planning. With regard to planning, the Secretary of State for Wales has broadly the same powers and responsibilities as the Secretary of State for the Environment in England. Since Wales shares the same legal system with England, planning legislation applies to both nations, and planning in both England and Wales is normally subject to the same guidance and circulars published jointly by the Department of the Environment and the Welsh Office – although, where appropriate, the Welsh Office issues separate guidance to take account of distinct Welsh circumstances. However, in an overview of strategic planning, the Assembly of Welsh Counties (1993) reported that the Secretary of State, rather than looking for the preparation of regional plans in the principality, had ensured that Planning Policy Guidance Note 15, *The Strategic Planning Guidance in Wales: structure plans and the content of development plans* (Welsh Office, 1990), was narrowly 'confined to those matters necessary to enable local planning authorities to prepare their structure plans and local plans' (Tewdwr-Jones, 1996: 41). Clearly, strategic planning guidance in Wales was of only limited value in facilitating the improvement of economic, social and environmental conditions on a regional scale. Less controversially, the Welsh Secretary – like the Secretary of State for the Environment – also determines planning applications that have been 'called in' for his decision, decides appeals against the refusal of planning consent and, where he considers intervention is necessary, calls in development plan proposals for approval. Like his counterpart in the Department of the Environment, the Secretary of State for Wales also has responsibility for the Planning Inspectorate, which became an executive agency in 1992.

At a sub-regional or regional level, planning has not evolved in Wales to the same extent as in England. Although the South Wales Regional Survey (Ministry of Health, 1921) is generally accepted as the first serious attempt at regional planning in the United Kingdom, over the following decades only two further plans were produced for areas of wide spatial extent – the South Wales and Monmouthshire Plan (Lloyd and Jackson, 1949), and Wales: the way ahead (Welsh Economic Planning Council, 1967). Although the latter plan was a product of machinery set up by the Department of Economic Affairs in 1965 to introduce regional planning across Great Britain through the medium of Regional Economic Planning Councils and Regional Economic Planning Boards (see p. 93), it was increasingly suggested that regional planning would only be effective if undertaken by elected regional assemblies (Mackintosh 1968). More substantial reasons for devolution were cited by the Kilbrandon Report (Royal Commission, 1973) in proposing that a directly elected assembly should be created in the principality to take over the functions administered by the Secretary of State. However, such a proposal was considered too far-reaching by the Conservative government of the day and consequently rejected. Opposition to devolution also seemed to be the view of the Welsh electorate in 1979, who voted overwhelmingly in a referendum against the Labour government's proposals to establish a Welsh assembly.

Although plans to devolve representative government had to be shelved, quangos were beginning to be set up in the 1970s – some with a degree of responsibility for regional planning in addition to their normal executive duties. Founded in 1975, the Welsh Development Agency (WDA) was committed from the outset to making indigenous enterprise the principal platform for the regeneration of the Welsh economy. To attract investment from both overseas and domestic sources, the WDA has – in collaboration with local authorities – invested funds in the reclamation of derelict land, developed several industrial parks, provided a full range of fully serviced and often redeveloped sites, purpose-built premises and advisory services, in addition to financing the formation and growth of firms. To assist with the provision of land for these and other purposes, the Land Authority for Wales (LAW) was also established in 1975 to take over the land acquisition and disposal role of local authorities. Whereas South Wales was often the location of WDA and LAW activity, much of the rest of the principality was the concern of another quango, the Development Board for Rural Wales (DBRW) – set up in 1977 to diversify the economic base of agricultural and tourist areas.

By the late 1990s, however, devolution was back on the political agenda. It was possible that the Welsh electorate was now more sympathetic to the idea of devolution – following a long period of minority Conservative government in Wales (and with no Conservative members returned at the general election of May 1 1997). A new Labour government therefore published a White Paper, *A Voice for Wales* (Welsh Office, 1997) setting out its proposals to create a sixty-member Welsh Assembly in 1999 with responsibility for broadly those matters dealt with by the Welsh Office in Cardiff. Although it would initially have an annual budget of £7 billion per annum, it would not have the ability to raise taxes or introduce primary legislation – its responsibilities being limited to secondary legislation which fleshes out the detail of primary legislation produced in London. Although it was the intention to scrap several of the forty-five main quangos, the role of the WDA would be strengthened by taking over some of the powers and resources of the demised quangos, notably the LAW and the DBRW. With 50.3 per cent of the poll in favour of the proposal at the referendum of 15 September 1997, the condition was met for the election to the assembly to take place on 6 May 1999. It can only be hoped that, in addition to implementing town and country planning legislation and other planning reponsibilities inherited from the Welsh Office, the new assembly facilitates the development of regional planning across the principality.

At a local authority level, devolution – in the short term – is unlikely to produce many changes to the gradually evolving system of town and country planning. As in England, the responsibility for producing development plans and for granting planning consent has changed over the years largely as a result of local government reform. The White Paper, *Local Government in Wales* (Welsh Office, 1967) proposed that in order to rationalise the provision of local services there should be five counties in Wales instead of thirteen, that Cardiff, Swansea and Newport should retain, but Merthyr Tydfil should lose county borough status, and that instead of 164 non-county boroughs and urban and rural district councils there should only be thirty-six new districts. The counties and county boroughs were the principal local planning authorities charged with the responsibility of producing development plans and granting planning consent, therefore reform along the above lines would have increased the spatial scale of physical planning. The Local Government Act of 1972 broadly accepted the rationale for change indicated in the White Paper and reduced the number of counties to eight and created thirty-seven new districts. Under the provisions of the Town and Country Planning Act of 1971, the county councils became responsible for producing structure plans, while the districts had responsibility for producing local plans. As in England, the preparation of structure plans in Wales was substantially influenced by the issue of guidance notes – PPG 12 (Wales) Development Plans and Strategic Planning Guidance in Wales (Welsh Office, 1992), for example, claiming – perhaps a little extravagantly – that structure plans 'have a key role in translating national and regional policies (Welsh Office policies) to a more tightly defined spatial area'. (Tewdwr-Jones, 1996: 57).

However, as a consequence of further local government reform, structure plans in the principality soon became obsolete. Under the Local Government (Wales) Act of 1994, twenty-two unitary authorities replaced the eight counties and thirty-seven districts created only twenty-two years earlier. Reform was considered necessary to restore authorities such as Cardiff, Swansea and Newport – cities which had been demoted to district status in 1972, and to re-create traditional counties such as Anglesey and Pembrokeshire which had been subsumed by larger counties in the same year. It was argued that the unitary system could be commended for its administrative simplicity (which could speed-up the plan-making process and enhance efficiency) and for being based on traditional geographical areas with which residents would be able to identify (Cullingworth and Nadin, 1993; Tewdwr-Jones, 1996).

The new unitary authorities have the responsibility for preparing unitary development plans – incorporating both strategic and detailed land-use policies. It is sometimes necessary for groups of authorities to collaborate in development plan process, but the Secretary of State was given default powers to intervene if strategic issues were not adequately addressed (Cullingworth and Nadin, 1993) – a role to be taken over by his devolved successor after 1999. Wales thus has a system of planning which differs markedly from that in non-metropolitan England, and, until the Welsh Assembly is a reality, the production and implementation of unitary development plans will be influenced by a strategic planning guidance exercise (initiated by the Welsh Office) which will be very different from regional planning guidance in England (Tewdwr-Jones, 1996). Clearly, after 1999, the Welsh Assembly will perform this role. It is important to recognise, however, that the new unitary authorities are spatially very small compared to regions elsewhere in the EU, and there is some doubt whether they will be able to submit coherent regional strategies as part of their bid for European assistance (Tewdwr-Jones, 1996). It can only be hoped that the Welsh Office or subsequently the Welsh Assembly will ensure that regional strategies are developed and applied on an appropriate scale.

Scotland

From the Act of Union in 1707 to 1999, Scotland was governed from Westminster through a succession of Secretaries of State for Scotland, although in 1885 the Scottish Office was created in an attempt to add substance to governance north of the border, but it was not until 1939 that it was re-located from London to Edinburgh.

The Scottish Office was responsible for broadly the same matters as the Welsh Office – health, community care, education, forestry and fisheries, economic development, urban policy, new towns, transport, local government, housing, water and sewerage, environmental protection, nature conservation and the countryside, and town and country planning. With regard to planning, the Secretary of State for Scotland, like the Environment Secretary in England and the Welsh Secretary, had the responsibility of implementing legislation governing the planning system; providing guidance and circulars on a range of policy issues as a means of assisting local planning authorities in the preparation and implementation of development plans; calling-in planning applications for approval; and involving its planning inspectors in appeal and inquiry work.

In a regional context, planning in Scotland dates back to the immediate post-war years. The Clyde Valley Regional Plan of Abercrombie and Matthew (1946) – a peak of achievement in its day – was a prelude to the production of a series of regional plans and surveys of varying quality throughout the following three decades. Physical plans were produced for Central and South East Scotland in 1948 and for the Tay Valley in 1949 but manifestly ignored such issues as the development of new towns and the relocation of overspill population. However, whereas the West Highland Survey of 1950 was – like surveys of earlier times – little more than a record of existing land uses, thirteen years later rising unemployment and political tension in the most populated area of Scotland resulted in a strong economic dimension being incorporated into 'Central Scotland: programme for development and growth' (Scottish Development Department, 1963). In the late 1960s and throughout most of the 1970s, numerous integrated regional reports were produced giving approximately equal weight to both physical and economic considerations. Plans were prepared by consultants for the Falkirk-Grangemouth growth area in 1968, North East Scotland in 1969, and the Borders in 1968, while the Scottish Development Department of the Scottish Office prepared plans for both Tayside and South West Scotland in 1970, and the West Central Scotland Plan of 1974 was prepared jointly by local authorities and the Scottish Development Department. However, whereas the central belt of Scotland had arguably the most varied experience of regional planning in Britain by the 1970s, the north and north-west of Scotland led the way in economic development on a regional scale. Set up in 1965, the Highland and Islands Development Board (HIDB) was successful in supporting indigenous economic activity (such as agriculture, forestry, fishing and tourism), attracting investment from outside of the region, and, in general, halting out-migration.

To counter the more radical aims of Scottish nationalism, and to achieve greater accountability not least within the broad field of planning and development, it was considered necessary to implement some form of devolved government north of the border. Based on three years of investigation and consultation, the Kilbrandon Report (Royal Commission, 1973) proposed that a directly elected assembly for Scotland should be set up to be responsible for those functions hitherto administered by the Secretary of State for Scotland (Wannop, 1995). The Conservative government of the day, however, rejected the Royal Commission's proposals, but, following the return of a Labour government in 1974, further plans for evolution were drawn up and put to the Scottish people in a referendum in 1979. Although there was a majority vote in favour of establishing a Scottish assembly, the majority was insufficiently large for appropriate legislation to be enacted, and plans to devolve representative government were shelved throughout the whole period of the subsequent Thatcher and Major governments, 1979–97.

During this period of Conservative administration, the Scottish Development Agency (SDA) (set up by the previous Labour government in 1975) performed a useful role in facilitating economic development, particularly in the central belt. Funded annually by the Exchequer, and under the control of the Scottish Office, the SDA reclaimed derelict land, rehabilitated the environment, built and managed industrial estates, leased or sold factories, invested in industry, established new companies, and provided industry with finance and advice. The SDA increasingly entered into partnership with both the private sector and predominantly Labour controlled-councils. It formed, for example, a venture-capital finance company with the privately owned Royal Bank of Scotland in the early 1980s, and in the period 1982–86 promoted Dundee as a centre of high-tech by investing – with Tayside region and the district council – £24 million locally in the expectation that the private sector would also invest substantially in the city. A similar emphasis on investment in modern industry has characterised the role of the SDA in helping to change the image of Clydeside from a region of heavy smokestack industry to one of electronics, energy-related technology and advanced engineering (Williams, 1985), and the SDA was also instrumental in planning and financing the Glasgow East Area Renewal (GEAR) project.

In 1991, the HIDB was replaced by Highlands and Islands Enterprises (HIE). Although the HIDB had increased the economic base of the north and north-west of Scotland, and in general had checked out-migration, there were nevertheless some areas where population loss was severe – for example in the Western Isles. The functions of HIE are therefore to run government-funded job-training schemes to supply a pool of skilled labour in the region and to support and oversee local enterprise companies (LOCs)

– private companies funded mainly by the government (Drake, 1994). Also in 1991, the SDA merged with the Training Agency in Scotland to form Scottish Enterprise, although most of the responsibilities of Scottish Enterprise were subsequently delegated to LECs.

By the late 1990s devolution returned to the political agenda. It was evident that a higher proportion of the Scottish electorate was now in favour of devolution following a long period of minority Conservative government, and the loss of all Conservative-held constituencies north of the border at the general election of 1 May 1997. A new Labour government therefore published a White Paper, *A Government for Scotland* (Scottish Office, 1997) setting out its proposals to establish a Scottish Parliament in 2000 with responsibility for broadly those matters dealt with by the Scottish Office in Edinburgh (including economic development and town planning). With a block grant of about £20 billion from Whitehall, it would have autonomy over spending, powers to introduce primary legislation, be able to increase or decrease the rate of income tax by 3p in the pound, be able – through its ministers – to deal directly with the European Union, and the Scottish Parliament would be authorised to scutinise EU legislation. With 74.3 per cent of the poll being in favour of devolution at the referendum of 11 September 1997, the condition was met for the election of a Scottish Parliament to take place in the first half of 1999.

At a more local level, constitutional reform had already occurred twice-over in the latter years of the twentieth century, first in the early 1970s and second in 1996. The Wheatley Report (Royal Commission, 1969b) – emphasising the need to establish an appropriate administrative structure to undertake planning on a regional scale – recommended that the existing local authorities (four cities, twenty-one large boroughs, 176 small boroughs and 196 districts) should be replaced by a two-tier system of seven regional authorities and thirty-seven district authorities. Refining this proposal, the Local Government (Scotland) Act of 1973 created nine regional councils, fifty-three district councils and three island authorities, but, of the total of sixty-five authorities, only forty-nine had planning powers, since several covered very large and barely-populated areas (Cullingworth and Nadin, 1994). Under the Secretary of State for Scotland, the responsibility for planning was shared between the regional and district authorities in the Central region, Fife, Grampian, Lothian, Tayside and Strathclyde, whereas in the Borders, Dumfries and Galloway, and the Highlands, planning was the exclusive responsibility of the regions. In Orkney, Shetland and the Western Isles, island authorities (rather than regions or districts) function as all-purpose planning authorities. Although the two-tier system was only applicable to the regions of the central belt, these contained nine-tenths of the population of Scotland, with Strathclyde alone having a population of over 2 million or 40 per cent of the total. Conceived as strategic authorities, the regions were responsible for planning on a regional scale, public transport, education, social work and water, whereas the districts were responsible for local planning, housing, refuse collection and disposal, and a range of other local services (Cullingworth and Nadin, 1994).

Like the English and Welsh counties, the Scottish regional and island authorities were required to produce structure plans to provide a strategic framework for the preparation of local plans – which, in turn, was a function of the district councils in the central belt regions. Until 1992, the regions were obliged to submit their structure plans to the Secretary of State for approval, but thereafter this procedure was continued. However, in contrast to England and Wales, local plan coverage became mandatory – possibly to compensate for it being inessential to secure full structure plan coverage. For reasons of practicality, the Scottish Development Department recommended that, in preparing local plans, 'priority should be given to those areas where development is expected and development pressures are likely to be greatest' (Thomson, 1985).

Whereas in England and Wales, the Environment and Welsh Secretaries began issuing Planning Policy Guidance notes in the 1980s to guide local authorities in the preparation and implementation of their development plans, north of the border the Secretary of State started issuing National Policy Guidelines (NPGs) in the 1970s on such matters as coastal planning and aggregate workings, and subsequently on such issues as priorities for development planning, skiing developments, high-tech land-use on sites of high amenity value, major retail development and agricultural land – all of which were intended to provide a framework for structure and local planning. However, in response to the consultative paper, *Review of Planning Guidance* (Scottish Office, 1991a), National Planning Policy Guidelines (NPPGs) replaced NPGs in an attempt to 'give greater clarity to strategic planning' (Lloyd and Black, 1995: 43).

As in Wales, the Conservative government in the early 1990s became committed to a unitary system of local government in Scotland, in contrast to England where the two-tier system has been largely retained. Based on two consultative papers, *The Structure of Local Government in Scotland: The Case for Change* (Scottish Office, 1991b) and *The Structure of Local Government in Scotland: Shaping the New Councils* (Scottish Office, 1992a), a single tier of twenty-eight unitary councils came into existence in 1996 replacing the former system of regional, district and island authorities. The responsibility for planning inevitably became fragmented – particularly in the cities of Glasgow, Edinburgh, Aberdeen and Dundee. In metropolitan Clydeside – where successive governments since 1944 had accepted the need for repetitive integrated strategic planning – the responsibility for planning, henceforth, was borne by at least ten independent councils (Wannop, 1995: 205). Strategic planning throughout much of Scotland was thus rendered impossible. The Conservative administration, ignoring the lessons of the previous sixty years, had, in effect, put the clock back to the 1930s. It might be assumed that the government believed that quangos such as Scottish Enterprise and Highlands and Islands Enterprise could be relied upon to provide the strategic framework necessary for planning at a local level, or it might have been thought that, because of easier travel and new commercial and social tendencies, regional boundaries no longer have any meaning (Wannop, 1995). An alternative explanation was that by abolishing the Scottish regional councils, the government would ensure that political power would be further centralised (a process akin to the abolition of the English metropolitan counties and the Greater London Council in 1986 – see p. 98).

Notwithstanding the recent demise of strategic planning, the Planning and Compensation Act of 1991 – as in England, Wales and in accordance with Planning Advice Note 37, (Scottish Office, 1992b) – aimed to enhance the status of development plans in development control, to bring about more succinct statements of policy, and to place an emphasis on physical land-use development (Cullingworth and Nadin, 1994), but in Scotland, in contrast to the rest of Great Britain, the structure plan still has to be approved by the Secretary of State – a further indication of centralised power.

As in Wales, it is necessary to recognise that the new unitary authorities are very small compared to regions in most other countries of the EU. It can only be hoped, therefore, that a devolved parliament not only reviews the current system of local government and town and country planning north of the border, but also considers how Scotland can again be divided into regions and how regional planning can be facilitated across the whole of the country. This will be essential if Scotland is to successfully bid for her rightful share of EU funds in the early twenty-first century.

Regional policy and regional incentives in Great Britain

From the 1930s to the late 1970s, regional policy was based on three premises: that the problem of economic depression in parts of the north and west of Britain was due to localised deficiencies in demand associated with the collapse of basic and staple industries; that intervention was necessary on social grounds to reduce regional imbalances in employment opportunities, and on economic grounds to utilise unemployed labour and thus facilitate a higher rate of non-inflationary growth; and that it was necessary – through a 'carrot and stick' policy of incentives and controls – to attract industry away from low unemployment growth regions to areas of high-unemployment (Martin and Tyler, 1992).

Under the Special Areas legislation of the 1930s and the Distribution of Industries Act of 1945, Development Areas were designated in the old industrial districts of, for example, West Cumberland, the North-East, South Wales and Central Scotland, but, by the mid-1960s, the Assisted Area status covered districts which contained 17 per cent of the working population of the United Kingdom. In 1966, the whole of the Northern region, Merseyside, the South West, large parts of Wales and all of Scotland were designated as Development Areas (or Special Development Areas where there were particularly high levels of unemployment), and from 1973 to 1982 the boundaries of the Assisted Areas expanded with the inclusion of new Intermediate Areas – the Assisted Areas now containing about 50 per cent of the United Kingdom's working population (Martin and Tyler, 1992).

In the post-war years, Industrial Development Certificate (IDC) controls – restricting new factory building in the South East and the Midlands – were accompanied by the construction of Government Advanced Factories and the reclamation of derelict land in the Development Areas. In the 1960s to 1970s, further policy instruments were introduced such as the Regional Development Grant (RDG), the Regional Employment Premium (available from the mid-1960s to the mid-1970s), and Regional Selective Assistance (RSA).

However, in the 1980s, the economic case for regional policy was rejected by the Secretaries of State for Trade and Industry, Scotland and Wales. Thatcher governments no longer believed that, through the medium of regional policy, employment growth and national output would be stimulated, but instead articulated in the White Paper, *Regional Industrial Development* (Department of the Environment, 1983) that the case for continuing with regional policy was 'principally a social one with the aim of reducing, on a stable long-term basis, regional imbalances in employment opportunities' (para. 16). Wedded, however, to the neo-liberal view that too much government intervention and a lack of enterprise were largely responsible for the predicament of the depressed regions, the government was already embarked upon a policy of rolling back the map of the Assisted Areas and attempting to alleviate supply-side weaknesses. From 1979 to 1982, Assisted Area coverage had been reduced from 44 to 26 per cent of the British working population, and in 1984, with the aim of simplifying regional policy and making it more selective, the government abolished the Special Development Areas, and tightened the boundaries of the Development Areas to contain only 15 per cent of Britain's working population – largely concentrated in the older conurbations, though the Intermediate Areas were expanded to cover approximately 20 per cent of the working population. In 1993, there were further revisions to the Assisted Areas map, but overall population coverage fell only slightly from 35 to 34 per cent (see Figure 6.3). IDC controls were discontinued in 1982 to remove any disincentive to develop in the South East and the Midlands, while, demand-side subsidies were increasingly reduced or discontinued. RDGs were reduced from 20 per cent to a uniform rate of 15 per cent in the Development Areas, with a cost limit of £10,000 for each new job created, or alternatively companies could claim employment grants of £3,000 per job. In 1988, RDGs were completely abolished, with regional aid being confined to the construction of Government Advanced Factories, and the provision of RSAs,

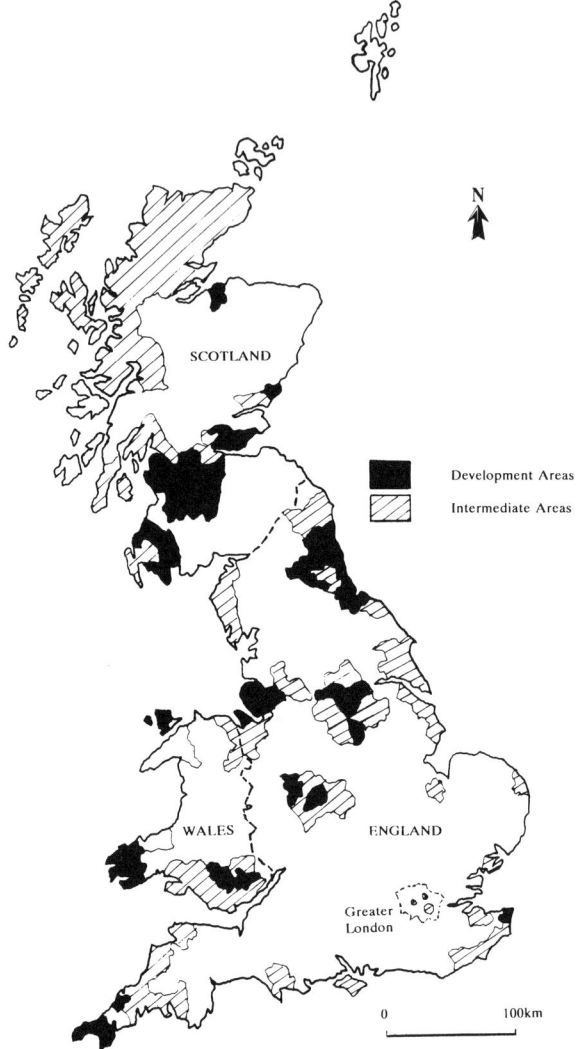

Figure 6.3 Government assisted regions, Great Britain, 1993

Regional Enterprise Grants (REGs) and regionally-differentiated Business Consultancy Initiatives targeted at small firms.

Northern Ireland

Following 51 years of devolved governments, a long period of direct rule by Westminster began in 1972. A Federal system of government under its own prime minister was replaced by a French-style *préfecture* in the guise of the Secretary of State for Northern Ireland. Planning in the province became the delegated responsibility of the Department of the Environment for Northern Ireland. In contrast to both Great Britain and the Republic of Ireland, planning in the north was not a responsibility of local government. Instead six divisional planning offices consulted with the 26 district councils and worked closely with them in the preparation and implementation of development plans. In general terms, the minister responsible for planning had the same powers as the Secretary of State for the Environment in England, although planning appeals could be made to an Independent Planning Appeals Commission.

Since 1921, regional policy in Northern Ireland has been distinct from the rest of the United Kingdom and is largely indistinguishable from industrial policy. Established in 1982 and based in Belfast, the Department of Economic Development has overall responsibility for industrial development policy in Northern Ireland. As elsewhere in the UK regional aid took the form of discretionary Selective Financial Assistance in the form of capital grants, employment grants, interest relief grants or soft loans, marketing grants and research and development grants.

As an outcome of the 1993–98 peace process, under a North-South ministerial council, cross-border bodies made up of representatives from a new Northern Ireland Assembly and the Dublin government will assume responsibility for matters of common concern, for example urban and rural development, the environment, transport and tourism.

Structural Fund allocations to the United Kingdom, 1994–99

For the period 1994–99, only Northern Ireland, Merseyside and the Highlands and Islands qualified for Objective 1 funding (see Figure 6.4). Through groups of counties (NUTS-2), they were respectively allocated 1,233, 816 and 311 million ECU to promote development and structural adjustment. By comparison, Objective 2 allocations were dispersed over a large number of areas including much of the central belt of Scotland, the north-east of England, West Cumbria, south-east Lancashire, southern Yorkshire and Humberside, the West Midlands conurbation, South Wales, south-west Devon and part of East London. Through the relevant local authority areas (NUTS-3), 4,581 million ECU were available in 1994–99, to facilitate the economic conversion of areas affected by industrial decline. With the exception of the lowland areas of East Anglia and Lincolnshire, Objective 5b areas are situated largely in 'Highland Britain' north and west of a line from the River Exe to the River Tees, and comprise Cornwall and north Devon, most of Wales, the Pennines, the Borders and parts of the Scottish Highlands. As vulnerable rural areas in need of economic diversification, they were allocated 817 million ECU for the five years, 1994–99. Areas eligible for EU regional aid contained, in total, 41.9 per cent of the population of the United

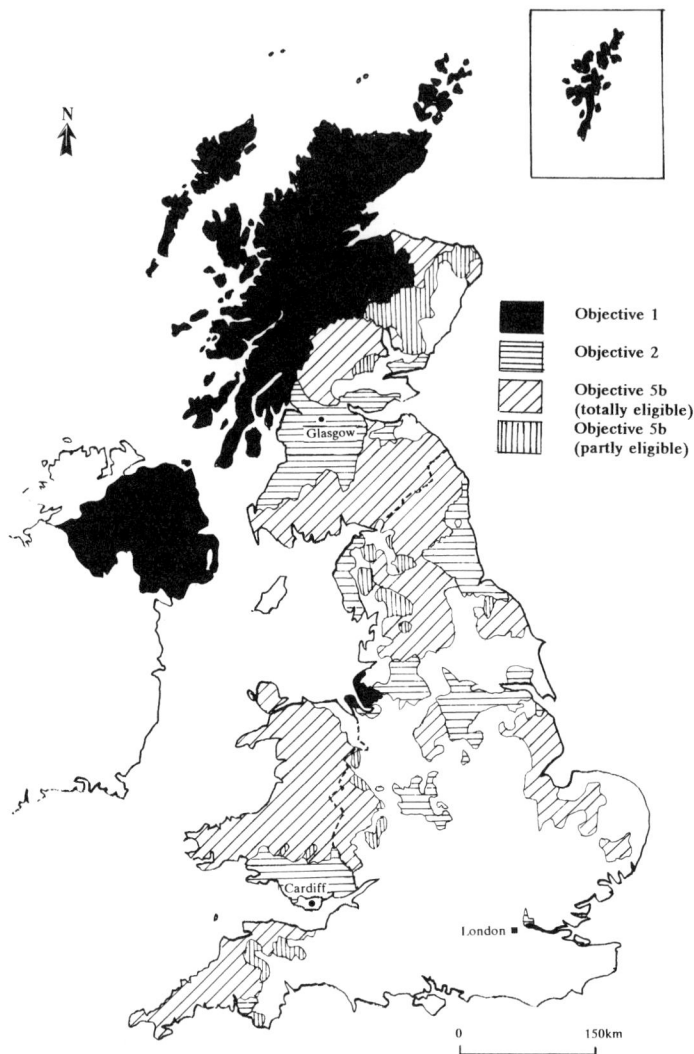

Figure 6.4 Designated Structural Fund areas, Great Britain, 1994–99

Kingdom, with 31 per cent living in Objective 2 areas, 6 per cent living in Objective 1 areas and only 4.9 per cent living in Objective 5b areas (CEC, 1996b).

The Netherlands

As is typical in much of Western Europe, the Netherlands has a three-tier system of government, with legislative and administrative powers vested in the national government, the twelve provinces and the 650 municipalities (*gemeenten*). Each level of government, in different ways, is responsible for planning (Bussard, 1986; Davies, 1989), but, perhaps more than in most other countries and reflecting the consensual nature of Dutch society, there is a considerable degree of both vertical co-ordination between the three levels of government and horizontal co-ordination between the different departments and agencies within each level of government – both obligatory and voluntary (Davies, 1989; Brussard, 1986; Newman and Thornley, 1996).

At the national level of government, the Ministry of Housing, Spatial Planning and the Environment (VROM) is responsible for preparing: the Physical Planning Act which provides the framework for planning throughout the Netherlands; statements of national planning policy which attempt to co-ordinate the activities of the various sectoral bodies; and occasional reports on physical planning in the Netherlands which set out national spatial planning priorities – the Fourth Report, *National Spatial Policy* (VROM, 1993) emphasising, for example, the need to strengthen the position of the country, and particularly the Randstad, within Europe, and to maintain the contrast between urban and rural areas, and to particularly preserve the *Green Heart* of the Randstad (EC, 1994).

The provinces, have extensive administrative powers due both to the principle of joint government (the provinces are represented at national government level by a Senate) and to the absence of any other intermediate tier of government dependent on them (Wiehler and Stumm, 1995). They are thus able to exercise an intermediary or co-ordinating role and exert their authority over the municipalities on matters such as physical and economic planning. It is at the level of the province that the general directives of each of the ministries are integrated and translated into orders to be implemented by the municipalities (Newman and Thornley, 1996). On this basis, each provincial authority produces a regional plan (*Streekplan*) setting out the strategic direction of land-use and transport across all or part of their province. The plan is administratively binding on the municipalities and, together with supplementary reports on specific topics, forms the basis for the approval of lower-tier plans (Brussard, 1995).

At the lower-tier of government, municipalities are blessed with a high degree of autonomy within their areas of responsibility. Within the context of planning, they have the right, though not the obligation, to produce a structure plan (*struktuurplan*), which is an indicative strategic document for all or part of the administrative area. They are thus free to decide their own strategy, 'provided that the policies do not conflict with the regional plan or policy statements made by the national government' (Ratcliffe and Stubbs, 1996). Although the structure plan does not need to be approved by a higher authority and is not legally binding, it provides a framework for the *Bestemmingsplan* (the local land-use plan) which a municipality is obliged to produce for the whole of its area. Subject to being approved by the relevant province, this type of plan shows the intended use of every parcel of land and enables the municipality to decide whether to grant building permits. As in other countries, where the legal system is based on the Napoleonic code, proposed development conforming with the statutory plan and building regulations cannot be refused, and non-conforming proposals cannot be approved (Newman and Thornley, 1996). There is nevertheless a degree of flexibility since the local land-use plan can be revised or withdrawn to permit development which the municipality considers, for example, worthy of approval or in keeping with a plan in preparation (Newman and Thornley, 1996: 49). Municipalities are also empowered to pursue a proactive role in acquiring and servicing land and in leasing or selling it to private developers or housing associations – a role which undoubtedly ensures that the interests of developers are taken into account when the *Bestemmingsplan* is being prepared (Newman and Thornley, 1996).

Since the responsibility for planning at municipal level is highly fragmented, it has become difficult to produce spatially integrated plans particularly in urban areas – despite the co-ordinating role of the provinces. In recent years, therefore, it became acknowledged that there was a need for a new tier of government in the Netherland's major urbanised area – the Randstad. In response to the report of the Montijin Commission of 1989 – which recommended the establishment of some form of metropolitan government – the *Kaderwet* of 1994 (the Framework Act) confirmed that it was the government's intention to promote the setting up of metropolitan government in the seven biggest urban areas. However, at the time of writing, none had been established, although it was anticipated that an elected government in the Rotterdam area would shortly be established to replace the Gemeente of Rotterdam and the Province of Zuid-Holland – facilitating more integrated planning of inner city and suburban areas.

Regional policy and regional incentives

In recent years, inter-regional policy in the Netherlands has aimed primarily to assist the development of parts of the north, Drenthe and South Limburg (areas which cover respectively 9, 3.8 and 2.8 per cent of the Dutch population). The main regional incentive, the Investment Premium (IPR) is a project-related capital grant and is available to manufacturing industry, 'footloose' services and tourist activities at a rate of 20 per cent of eligible investment. In addition, under the Space for Economic Activity Stimulation Scheme (StiREA), projects designed for the development, access and restructuring of business locations within a regional context are eligible for assistance; while the Integrated Structural Programme for the North of the Netherlands (ISP) helps to reinforce the private sector, and improve the infrastructure and the regional business environment northern development area (Yuill *et al.*, 1996). From 1995 to 1999, the Ministry of Economic Affairs plans to spend Fl 508 million on IPR grants, Fl 300

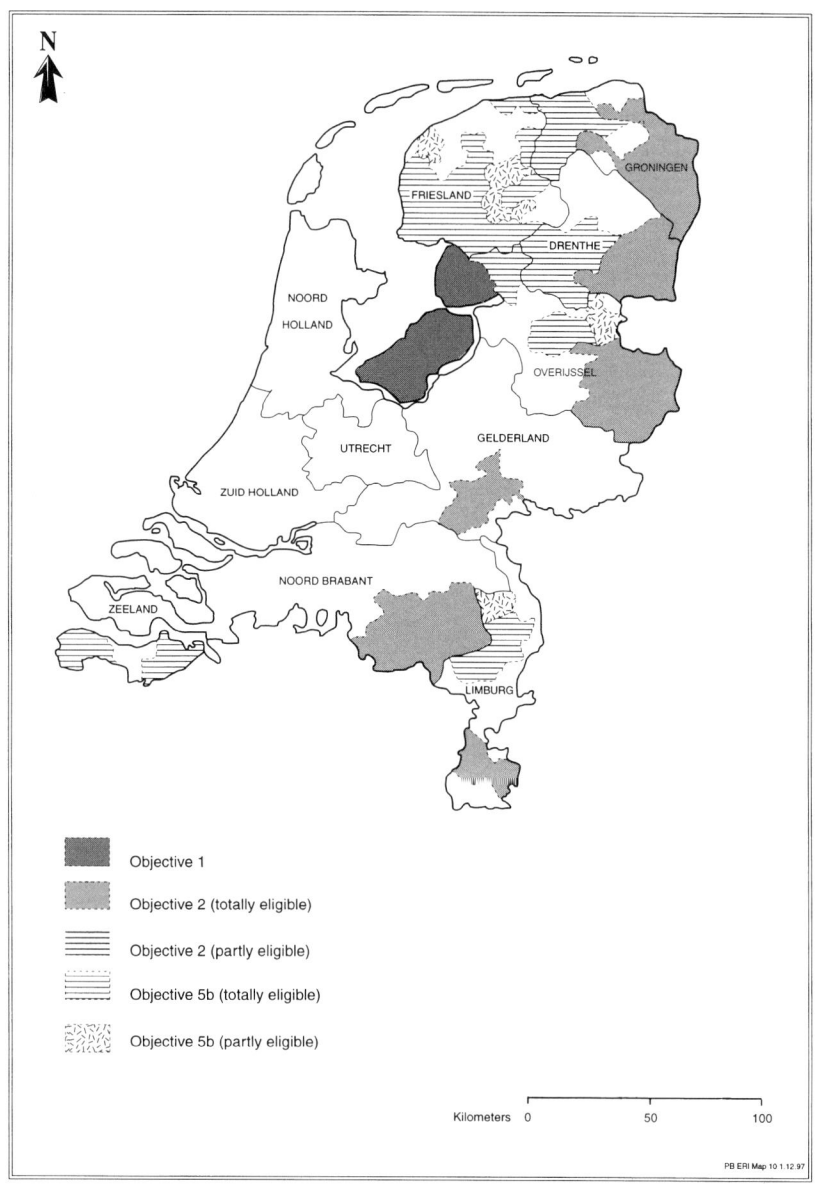

Figure 6.5 Designated Structural Fund areas, Netherlands, 1994–99

million on StiREA assistance and Fl 360 million on ISP aid – in total equivalent to about 560 million ECU.

The extent of inter-regional aid has been markedly reduced since the early 1980s. In 1982, IPR areas contained 28.7 per cent of the population (compared to only 15.6 per cent in 1995), while the maximum award was as much as 35 per cent of eligible investment in the worst-off localities. By 1988, IPR Area coverage had been reduced to 19.9 per cent and the maximum award rate had been cut back to 25 per cent. These reductions, in part, can be attributable to the adoption of neo-liberal policies by successive governments and, in part, due to greater reliance on EU funding.

With the allocation of Structural Fund for 1994–99, only one NUTS-2 area, Flevoland was sufficiently lagging behind to be eligible for Objective 1 funding (see Figure 6.5). A sum of 150 million ECU became available to facilitate the development and structural adjustment. A single programming document was consequently produced analysing the economic situation of the area and containing a set of priorities to which proportions of the budget were allocated. Flevoland consists of reclaimed polders and contains a very efficient agricultural sector currently facing problems owing to the fall in agricultural prices, while its newly developed urban areas accommodate overspill from the Randstad (Williams, 1996). Taking this situation into account, it was decided to prioritise agriculture, agribusiness and rural development; the regional business sector, in particular small and medium-sized enterprises; tourism; and the fisheries sector. Parts of the provinces of Drenthe, Friesland, Gronigen, Overijssel, Noord Brabant and Limburg (all NUTS-2 areas) were designated as declining industrial areas and were thus eligible for Objective 2 funding. A sum of 650 million ECU was thus made available to these areas to help facilitate their economic conversion. Objective 5b areas were confined to the northern provinces and parts of Limburg and Zeeland, and received funding of 150 million ECU to promote alternative activities to agriculture. To secure funds for their eligible areas, provincial governments are obliged to submit their plans to the national government for consideration before an application for funding is submitted to the European Commission.

Areas eligible for Objective 1 funding contained only 1.45 per cent of the total population of the Netherlands in 1994, Objective 2 areas contained 17.3 per cent, while 5.4 per cent of the population lived in Objective 5b areas (CEC, 1996b).

France

In France, arguably more than elsewhere in Western Europe, it has been accepted across most of the political spectrum since the Second World War that state planning is central to economic development. Every five years, a national development plan not only helps to determine economic policy but also influence spatial planning. For this reason, and because of the constitutional need to maintain equality of opportunity throughout the country, the planning system in France until recently was highly centralised. Central government had overall competence for the production of land-use plans, *Plans d'Occupation des Sols* (POLs) and for the control of development, processes which were applied in each of the *départements* through the medium of local offices of the planning ministry, the *Département de l'Equipement* (DDE). Within each *département*, a centrally appointed *préfet* (with his/her technical experts) had the responsibility of approving the POL and for issuing building permits to compliant developers. Since 1982, the state has shared its responsibilities for planning with twenty-two regions, ninety-five *départements* and 36,433 communes, France changing rapidly from 'a system of centrally dominated administration to one of a plurality of competing levels of government' (Smith, 1997). While each level of administration is run by councils of elected representatives, only the state has legislative power, and regional and local government is required to observe the 1957 (but much amended) Code de l'Urbanisme et de l'Habitat which specifies authorised types of planning activity and regulations concerning the development of land (Newman and Thornley, 1996). The central state, moreover, is advised from time to time on spatial issues by DATAR, the *Délégation à l'Aménagement du Territoire et à l'Action Régionale*, a regional arm of government which, since 1963, has co-ordinated the activity of regional agencies and allocated regional development funds.

The central state

The state has overall responsibility for major infrastructural development, for establishing and maintaining the rules and guidelines relating to planning, and for periodically reviewing the planning competence of regional and local government. With regard to infrastructure, in the 1960s and 1970s the state assumed responsibility for planning large-scale housing estates within designated ZUPs (*Zones à Urbaniser en Priorité*), and in the 1980s and 1990s it determined the location and form of a number of *grands projects*, largely office and cultural development in the Ile de France (the Paris region), and also established much of the *Train à Grande Vitesse* (TGV) network. In respect of planning practice and procedure, by its deployment of *préfets* in the regions and local government areas and by its distribution of ministerial offices in most cities and large towns and cities, the state has attempted to ensure that it maintains control over the use and development of land nationwide (Newman and Thornley, 1996).

The regions

The newly constituted regional councils (*conseils régionaux*) co-ordinate economic development and, in consultation with the state, set out planning agreements in respect of transportation projects and other major public investment schemes. The regions are also authorised to prepare development plans (*aménagement du territoire*), but to date few have done so partly because there is continuing debate about the formation of larger regional units. They also have responsibility for the environment (in particular regional national parks). As Figure 6.6 shows, the regions contain two to eight *départements*, with populations varying in size from less than 1 million in Limousin to over 9 million in the Ile de France (Wannop, 1995). Regional planning in the Ile de France, however, is the responsibility of the *préfet* mainly because the state wishes to maintain direct control over the Paris region as the political and economic hub of the nation.

Established initially in 1964 for economic planning purposes, the regions, from their inception, became subject to a substantial degree of

Figure 6.6 Conseils régionaux and départements of France

state intervention. In 1964, regional *préfets* became responsible for preparing five-year economic development programmes, assisted by officials of the *commission à économiques* and supported by a *conference administrative régionale* and a *commission de développement économiques régionale* (comparable to the Regional Economic Planning Boards and the Regional Economic Planning Councils in England in the 1960s and 1970s) (Wannop, 1995), and in the 1970s central government set up Economic and Social Committees in the regions to reflect and promote the interests of both government and private interests – a role which continued after decentralisation. Since the mid-1980s, moreover, *conseils régionaux*, in close co-operation with *préfets* (and at a later stage with state enterprises, *départements* and communes), co-ordinate major investment schemes within the regions, produce *contrats des plan* and share the cost of implementation (Newman and Thornley, 1996). Successive contracts in 1984–88, 1989–93 and 1994–98 have focused on the issue of regional imbalance. Between 1989–93 contracts emphasised job-creating investment in areas of high unemployment (Chicoye, 1992), and between 1994–98 contracts favoured the poorer regions of Corse and Nord-Pas-de-Calais in terms of state expenditure at the expense of the more prosperous regions such as the Ile de France. *Contrats de plan* have resulted in the state, rather than the regions, incurring the greater share of expenditure on regional development. Between 1989 and 1993, for example, the state spent FF48 billion whereas the regions spent FF40 million (Drake, 1994). The state further advanced its influence over development within the regions by decentralising government departments and public agencies from Paris to major urban areas throughout France. Despite the introduction of elected *conseils régionaux*, there is thus only a very low degree of regional autonomy. Clearly, 'decentralisation at regional level represents only a partial devolution of power' (Newman and Thornley, 1996: 158).

The influence of *conseils régionaux* is also reduced by the ability of large cities to dominate regional economies, particularly if the cities form alliances with central government and the *départements* (Biarez, 1993). Ironically, when a city becomes a location for new administrative buildings and council chambers, it is often the city which benefits most in economic and political terms rather than the region (Newman and Thornley, 1996). It is thus argued that 'though French regions were established as functioning political and administrative institutions, overall they play a minor part in the policy-making process' (Le Galès and John, 1997: 51). The total budget of the average region is often no bigger than that of the regional capital city or average *département*.

Notwithstanding the processes by which the French constitution limits the power of the *conseils régionaux* to act independently within the planning arena, the regions are vigorously attempting to promote themselves within the European Union. Individual regions are recipients of EU funding, while Languedoc-Roussillon and Midi-Pyrénées have formed a Euroregion with Cataluña, and Nord-Pas-de-Calais have established a similar linkage with

Bruxelles, Vlaams Gewest, Region Wallonne and Kent (Condaminès, 1993). In general, however, the *départements* are often more effective claimants to EU funding than are the regions.

The départements

It is remarkable that the creation of the regional tier of government in 1992 'went hand in hand with the strengthening of the *départements*', (Le Galès and John, 1997: 53) thereby reinforcing their role as the core of the French administrative system. Although planning responsibilities were devolved to the regions and communes, and direct land-use responsibilities were transferred to the lower tier of government,' 'the influence of the "départements" remained undiminished in these areas. By becoming recipients and hence distributors of important social budgets and being able to offer technical advice to the smaller communes, they had an overarching impact on spatial planning and development control at the local level (Le Galè's and John, 1997; Newman and Thornley, 1996). The *départements* moreover, and not the regions, had direct responsibility for rural development (including rural infrastructure and local development), and environmental amenities such as footpaths.

With regard to EU funding, the *départements* were often in a stronger position than the regions. In the aftermath of the Single European Act of 1986, the distribution of structural funds owed their size more to the pre-existence of relationships between local actors and the Commission than to the "objectivity" of the selection criteria used' (Smith, 1997: 121). Invariably, operational programmes drawn up at a local level in 1989 'were more frequently dominated by *département*-led priorities than by a supposedly more "strategic" level of regional government' (Smith, 1997: 121), although in the 1990s, the central state, through the medium of the *Secretariat Général d'Action Régionale*, often assumed a dominant role in policy-making despite the regions and 'départements' contributing a greater share of Structural Fund co-financing.

The communes

Under the *Loi d'Orientation Foncière 1967*, a two-tier system of plans was introduced and was initially applied by central government before becoming a responsibility of the communes in 1982. The upper or strategic tier of planning, however, was not obligatory, and by 1990 only 200 *Schéma Directeurs* (SDs) had been produced (Acosta and Renard, 1991). Although a SD can be produced by a single commune, it must receive the approval of the majority of communes in the area affected by the plan, but, since the structure of local government is fragmented, strategic planning has generally become an intercommunal responsibility undertaken in partnership with the state (Newman and Thornley, 1996). The strategic plan must take full account of the programmes of the central state and other public bodies, and focus on general economic and social developments and major infrastructure needs in the areas concerned. The lower-tier of planning is almost entirely the responsibility of the communes, although the state, other public bodies and chambers of commerce are normally involved in the preparation of plans, and there are opportunities for public consultation and public hearings for objectors. This contrasts with the top-down system applied in the United Kingdom and Germany which necessitates conformity with plans and policy produced by the higher tier authority (Ratcliffe and Stubbs, 1996).

Each commune with a population of 50,000 or more is authorised to produce a *Plan d'Occupation des Sols* (POS). Whilst the production of a plan is not obligatory, communes are empowered to use a POS to control the development of land and to provide a guide for the acquisition of key development sites. Apart from indicating the socio-economic attributes of the area, a POS is essentially concerned with land-use zoning, dividing land into: 'U' zones (land which is already urbanised and where development will be normally allowed); 'NC' zones (mainly agricultural land); 'ND' zones (conservation areas) and 'NA' zones (future growth areas). Where an authorised commune fails to produce a POS, the state might deem it necessary to use its comprehensive powers to control the appearance and location of buildings (Newman and Thornley, 1996). Communes have also been given the responsibility for controlling development in sensitive areas such as historic town centres, mountains and coastlines, and for declaring, in areas of economic depression, *Zones d'Aménagement Concerté* (ZAC) in which planning permission is automatically granted for a range of joint public/private urban regeneration schemes (Ratcliffe and Stubbs, 1996).

Under planning law, communes have the ability to intervene in the land market to ensure that planning objectives are secured and, as part of this process, prevent land from being withheld from the market for speculative reasons. Since vendors are required to notify the communes of impending land sales, the communes frequently buy land by agreement at market value or, in other cases, communes might need to exercise their considerable compulsory purchase powers to acquire land for public purposes (Newman and Thornley, 1996). Communes might also require developers to contribute to infrastructure costs, while a charge might be imposed on development rights about a standard level. If an application for planning and building permission conforms with the POS or ZAC plan, the mayor of the relevant commune issues a *permis construire* to the developer, although in parts of France, particularly in the south, much development is illegal. Where communes follow illegal procedures, the state intervenes through the medium of *préfets*, whilst financial probity is scrutinised by regional *Cour des Comptes* (Newman and Thornley, 1996).

Since the majority of communes have very small populations (80 per cent have less than 1000 inhabitants), are small in area and have limited resources, it is both inappropriate and impossible for them to exercise their planning and development control powers, and are therefore still dependent upon the DDE for these activities. Whereas many of the smaller communes have formed *syndicats* to enable them to exercise their planning functions, in Bordeaux, Lille, Lyons and Strasbourg, *communautés urbaines* (CUs) were mandatorily established in 1961 to co-ordinate planning and infrastructural development across their many constituent communes. Subsequently, and for the same purpose, CUs were set up by local choice in Brest, Dunkirk, Le Creusot–Monceau-les-Mines, Cherbourg and Le Mans (Wannop, 1995). Outside of the conurbations, the larger communes or groups of communes often formed partnerships with the state, *Agences d'Urbanisme*, to jointly produce POS and ZAC plans and control development.

At the local level, development control has become far more flexible in recent years, particularly during the period of the centre-right government, 1993–97. Communes, to a varying extent, have attempted to attract new businesses to increase local tax revenue in response to a reduction in the value of block grants from central government. Concern has been expressed that, in the context of planning, communes are thus competing rather than co-operating with each other, and that there has consequently been a 'loss of some of the perceived virtues of the formerly centralised system such as equality of treatment and the consistent application of rules' (Newman and Thornley, 1996: 174).

Regional imbalance and regional incentive policy

Since the Second World War, an increasing proportion of the population and economic activity of France has become concentrated in the larger urban areas and particularly in Paris. In 1960, whereas the capital occupied only 2 per cent of the area of France, it contained 19 per cent of its population and 29 per cent of its industrial employment (Hall, 1992). In contrast, the rural areas suffered stagnation and decay, most notably to the west of a line drawn from Cherbourg to Marseilles, while the old industrial regions of, for example, Nord-Pas-de-Calais and Lorraine suffered economic depression. Although constitutional reforms in 1982 were intended to devolve numerous government functions to new *conseils régionaux* and the communes to constrain the overwhelming political and economic influence of Paris, more than two decades earlier, under the direction of DATAR, a planning framework was introduced to curb the economic and spatial growth of Paris, to establish alternative areas of growth, and to regenerate the older industrial areas of the north and east.

In 1960, to stop the future physical growth of the Paris region (the Ile de France), a ten-year plan, the *Plan d'aménagement et d'organisation général* (PADOG), was approved by central government. A permanent planning body, under a *Délégué-Général* (DG) working in liaison with a consultative regional assembly and a representative board of local and central government, was responsible for allocating budgetary funds to selected projects. It was soon accepted, however, that the plan was doomed to failure since it was applied to too small an area, and was based on the unrealistic assumption that the rate of in-migration would halve and that the capital's physical growth would be contained. The plan was thus substantially revised by the DG in 1965 with the introduction of the *Schéma Directeur d'Aménagement et d'Urbanisme de la Région de Paris*. For the next twenty-five years, the *Schéma* was based on the assumption that Paris would remain an expanding capital, and focused on three major development programmes: the reinvigoration of inner suburban zones, most notably 'La Defense' – a commercial, cultural and administrative growth centre to the west of the central area of Paris; the construction of two north-south rapid suburban railway links and three ring motorways; and the development of five large new towns (St Quentin-en-Yvelines, Melun-Sénart, Marne La Vallée, Cergy Pointoise and Evry) on north-west, south-east axes either side of the Seine.

To provide counterweights to the predicted growth of Paris, eight *métropoles d'equilibre* were designated in 1965. The counterweights, Lille-Dunkerque, Nantes-St Nazaire, Bordeaux, Toulouse, Marseilles-Aix, Lyons-St Etienne-Grenoble, Strasbourg and Nancy-Metz were selected because they were all major concentrations of urban population located some distance away from the immediate influence of Paris. Except for Bordeaux, Toulouse and Strasbourg (where existing resources could be put to use), it was necessary to set up special teams, *organismes d'aménagement d'aires* (OREAM) to plan the economic and spatial growth of these cities (Wannop, 1995), while, at the same time, central government provided incentives for economic development outside of the Paris region. The state invested in infrastructure throughout France and attempted to strengthen the economic base of eight of the twenty-two regional capitals and co-ordinate growth and change in all regions outside of the Ile de France (the subject of the *Schéma Directeur*) (Pallard, 1993).

As a further means of attempting to reduce regional imbalance, the state set up a number of agencies to co-ordinate planning in rural areas. Funds were provided for *remembrement*: the rationalisation of farms into larger and more productive units to improve the use of farming techniques, to ease the development of co-operatives and to facilitate more efficient marketing. Tourism was also encouraged, national and regional parks were established, and Brittany and the Central Massif became areas of 'rural renovation' (Minshull, 1996). With regard to industrial imbalance, conversion grants were targeted at the older coal-mining, steel and textile areas and helped to facilitate the relocation of industry from Paris to cities such as Clermond Ferrand, Nantes, Rennes and Toulouse. In addition, fifteen special development poles were set up to attract investment into the most depressed industrial areas such as Nord-Pas de Calais and Lorraine (Minshull, 1996).

Responsibility for co-ordinating the emerging system of regional planning was vested in a number of state agencies, DATAR, the *Commissariat*

Général du Plan and the *Commission Nationale d'Aménagement du Territoire* (the *Commission* being absorbed into the *Commissariat* in 1967). Later, in 1972, unelected representatives of the *départements* and communes were invited to participate in regional planning solely in an advisory role and were not permitted to employ resources or undertake expenditure on their own initiative (Wannop, 1995). Over the following twenty years, it became clear that the *métropoles d'Equilibre* were unable to attract sufficient investment and economic activity to enable them to be successful counterweights to Paris. Neither were these cities successful in spreading investment throughout their thinly populated rural hinterlands (Hall, 1992). If the historic distinction between the centre and the periphery has become eroded in recent years, this may have been due more to improved transport and communications, and to telecommunication and computer links, than to the planned development of regional cities or the availability of grants for rural development or industrial conversion.

Regional planning and the Single European Market

In the context of the Single European Act of 1986 and the subsequent formation of the European Single Market, both the central government and the regions were concerned about the competitiveness of the French economy (Chicoye, 1992) and recognised the urgency of adjusting territorial policy to enable both the cities and regions of France to compete more effectively in Europe.

In terms of introducing policy to enhance competitive ability, the capital and its region very quickly took the lead. Because the population of the Ile de France planning region had unpredictably stabilised at 10.7 million by the end of the 1980s as a result of the economic and demographic slowdown in France throughout the decade (Minshull, 1996), planning aims and objectives required revision. The *Schéma Directeur* was thus revised by the *Livre Blanc du L'Ile de France 2000* of 1990 which was 'designed to enable Paris to compete successfully with other EU capitals' (Minshull, 1996: 364). Although it was a policy aim to limit the region's population to 12 million by 2015, five major growth poles were to be established within the region to concentrate development: two at the edge of the Paris agglomeration (an international pole at Rossy-Charles de Gaulle airport in the north and a technopole at Saclay-Massy in the south-west), and three around existing centres in the middle ring at La Défense-Gennevilliers-Montasson in the west, the Upper Seine valley in the south-east, and at La Villette and La Pleine Saint-Denis in the north (Hall, 1992; Minshull, 1996).

The Guichard Report of 1986, however, had recognised that whereas Paris could undoubtedly compete with rival European regions, many other parts of France were significantly less competitive. The report therefore proposed that twelve urban areas outside of the immediate influence of the capital should be selected as the key beneficiaries of regional aid

(Gichard, 1986). DATAR, under the political control and direction of the ministerial *Comité Interministériel pour l'Aménagement du Territoire*, was given the new responsibility of increasing the European competitiveness of the regions and principal cities, and a *Charte d'Objetif* was introduced in 1991 to involve the largest cities and regional *préfets* in the process of identifying European-scale projects. *Contrats de plan* would subsequently be revised to take account of investment in European projects, while *préfets* would relate centrally provided services to regional priorities (Newman and Thornley, 1996).

Although the constitutional reforms of the 1980s remained intact into the 1990s, it was becoming increasingly clear that the structure of the administrative regions were neither fully compatible with the spatial economy of France nor as politically autonomous as the regions of some other countries in the EU. There was also growing concern that the smaller regions would be unable to compete with their larger European rivals. In response to a DATAR analysis of the late 1980s, which suggested that on planning grounds there was a need to create seven large regions defined by their proximity to the core of the European economy (the Milan-Frankfurt-London axis), the law on the Territorial Administration of the Republic 1992 prompted the setting up of a number of inter-regional bodies to consider the formation of these new units. In response to a DATAR publication, *Le Livre Blanc du Bassin Parisien*, 1992, and as an outcome of the joint ministerial/DATAR *Charte du Bassin Parisien* of 1994, the eight regions of the Paris Basin (Bourgogne, Centre, Champagne-Ardenne, Basse-Normandie, Haute Normandie, Pays de Loire, Picardie and the Ile de France) produced a joint *contrat de plan* which, in effect, absorbed the Ile de France with the aim of diverting much of its growth throughout a new potentially highly competitive European region (Newman and Thornley, 1996). Within the context of the *Charte du Bassin Parisien*, a new *contrat de plan* for the Ile de France was negotiated and the resulting *Schéma Directeur* scaled down the region's population and employment targets and restricted the amount of land to be developed (Newman and Thornley, 1996).

Less emphasis was placed on further constitutional reform and the continuing dominance of the central state. It is unlikely, moreover, that the eventual grouping of the twenty-two regions into seven planning areas will lead to greater autonomy. Although the national development plan, 1994–98, intended to decentralise responsibility for either the customs or housing administrative services to a new north region, consumer affairs to Loire-Armorique and the judiciary to the south-west, regional representatives recognise that the relocation of the services of the state is not the same as devolution and fear that *préfets* will have an enhanced role rather than a diminished role in planning (Newman and Thornley, 1996). The extended power of the state was also directed at local government. Because more than 36,000 communes have been very slow to group themselves into larger units (or communities of communes), a further tier of local government was proposed after the presidential elections of 1995. Between 350 and 450 'pays' would be established within the hierarchy of administration between the *départements* and communes and corresponding in scale with existing *arrondissements* (Scargill, 1996). Since the *pays*, to a great extent, would be economically cohesive they would form the basis for economic development and planning. Since each *pays* would be the responsibility of a *préfet*, there would be absence of local autonomy, while the *départements* (having lost some of their powers in 1982) feared a further reduction in their role, and like other levels of administration 'were concerned about growing prefectural powers' (Newman and Thornley, 1996: 173).

Clearly, since the Second World War, central governments in France have developed a highly complex system of decentralised planning within a reformed system of territorial administration. Although policy has aimed at correcting regional imbalance and subsequently at enhancing the ability of the regions and cities to compete in Europe, through maintaining or extending the role of the *préfet* the state has singularly failed to devolve legislative power away from Paris.

Regional policy and regional incentives

In 1982, not only were there major constitutional changes but there were also substantial revisions to the French system of regional incentives mainly through a process of rationalisation. A new regional policy grant (PAT – the *prime d'aménagement du territoire*) and a regional employment grant replaced a total of five separate schemes; and whereas previously there were three grades of recipient areas, the number was now reduced to two. The maximum grant of 25 per cent, however, remained unchanged, although in absolute terms the amount of aid per job created was broadly doubled. There was also a degree of decentralisation in the administration of grants in respect of small and straightforward applications) (Yuill *et al.*, 1997).

However, major cuts in public expenditure between 1984 and 1987 resulted in PAT being transformed from a semi-automatic grant for *investment* in problem regions to a discretionary grant aimed at influencing *location* decisions, and from January 1987 all applications were dealt with centrally to ensure a more effective control over distribution (Yuill *et al.*, 1997).

In 1990, the European Commission agreed that a number of areas which had hitherto received assistance under exceptional provisions could now be designated as permanent problem regions, and in 1991 two minor schemes were introduced, the first to encourage relocation away from the Ile de France, and the second to provide an incentive to develop smaller projects in rural areas (the latter being discontinued in 1993).

In 1995, as an outcome of a major national debate on regional policy, a new Framework Law for Regional Development (*loi d'orientation pour l'aménagement du territoire et le développement*) set out the aims of regional development policy to the year 2015. It increased the size of the regional policy

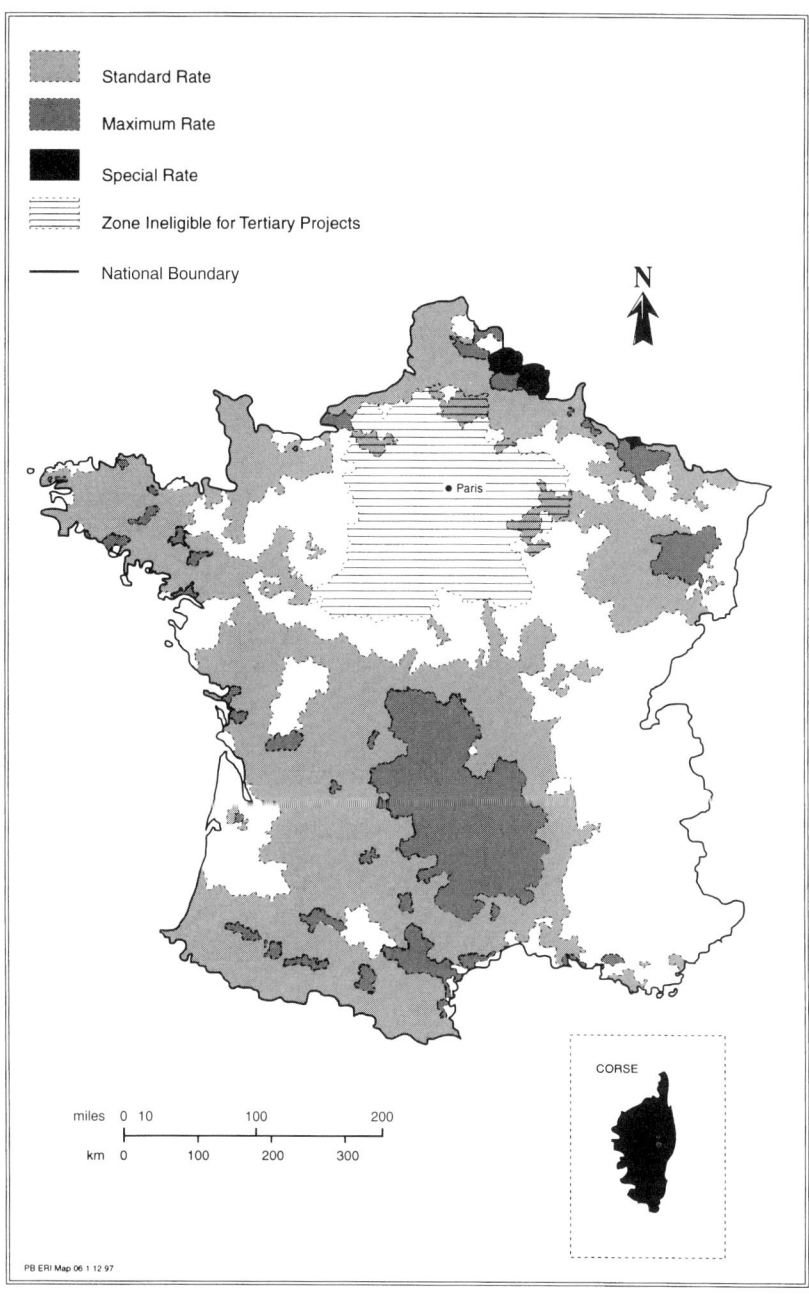

Standard Rate

Maximum Rate

Special Rate

Zone Ineligible for Tertiary Projects

National Boundary

N

• Paris

CORSE

miles 0 10 100 200

km 0 100 200 300

PB ERI Map 06 1 12 97

Figure 6.7 Government assisted regions, France, 1996

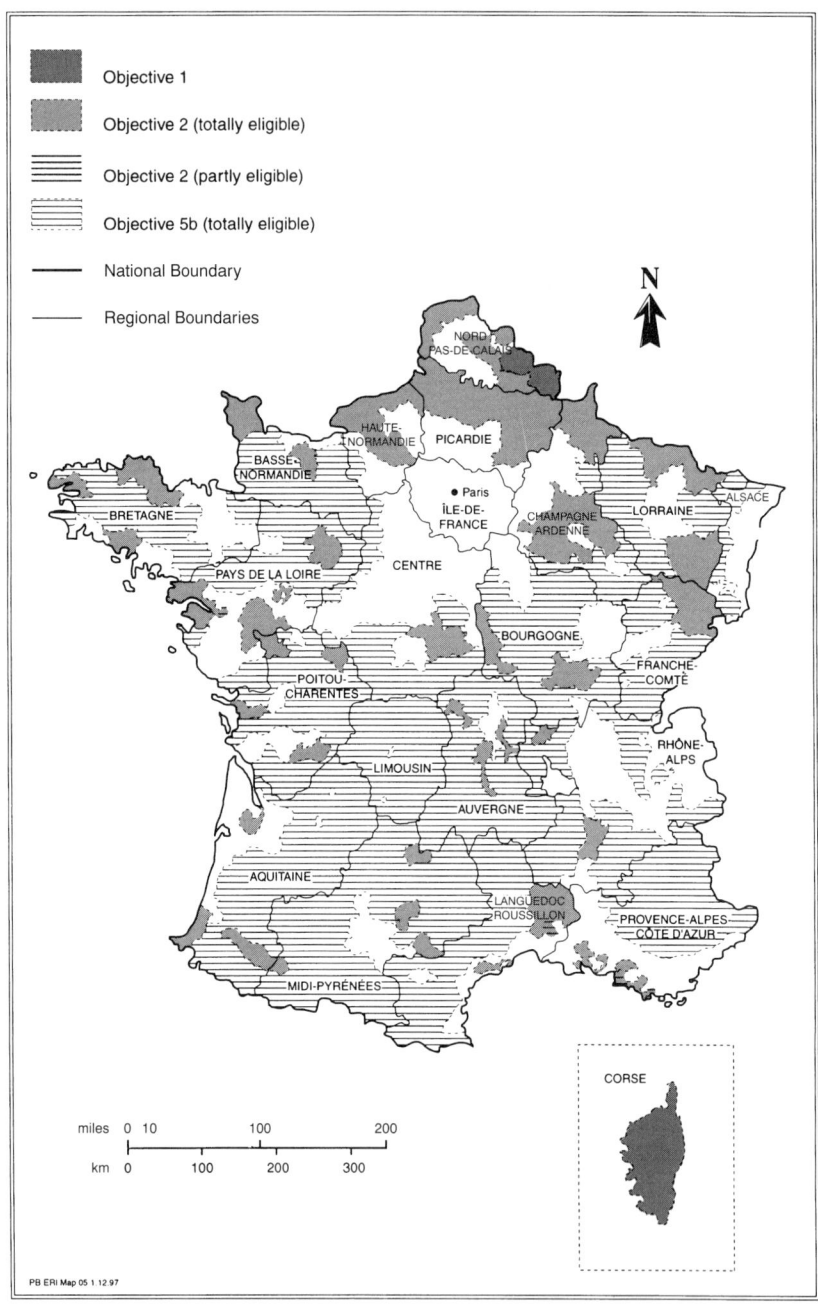

Legend:

- Objective 1
- Objective 2 (totally eligible)
- Objective 2 (partly eligible)
- Objective 5b (totally eligible)
- National Boundary
- Regional Boundaries

N

NORD-PAS-DE-CALAIS

HAUTE-NORMANDIE   PICARDIE

BASSE-NORMANDIE

● Paris
ÎLE-DE-FRANCE

BRETAGNE

CHAMPAGNE-ARDENNE   LORRAINE   ALSACE

PAYS DE LA LOIRE

CENTRE

BOURGOGNE

FRANCHE-COMTÉ

POITOU-CHARENTES

RHÔNE-ALPS

LIMOUSIN

AUVERGNE

AQUITAINE

LANGUEDOC-ROUSSILLON   PROVENCE-ALPES-CÔTE D'AZUR

MIDI-PYRÉNÉES

CORSE

miles   0 10   100   200
km   0   100   200   300

PB ERI Map 05 1.12.97

Figure 6.8 Designated Structural Fund areas, France, 1994–99

grant (PAT) in relation to job guidelines. PAT henceforth varied from FF50,000 (7,800 ECU) per job created in the standard award zones to FF70,000 (10,900 ECU) per job created in the maximum rate zones (see Figure 6.7); or, in respect of manufacturing projects, from 17 per cent of eligible expenditure in the standard award zones to 25 per cent in the maximum rate zones, 28 per cent in the Objective 1 area of Nord-Pas-de-Calais and 34 per cent in the Lorraine Development Pole and in Corse. The area of the country eligible for PAT was, however, marginally reduced from 41.9 per cent to 40.9 per cent of the national population.

When the distribution of EU Structural Funds for the period 1994–99 was decided, it was only part of the Nord-Pas de Calais region (in the *arrondissements* of Avesne, Douai and Valenciennes) and Corse which qualified for the whole of the country's Objective 1 allocation (see Figure 6.8). At the regional level of government (NUTS-2), a sum of 2.2 billion ECU was available to promote development and structural adjustment. In contrast, Objective 2 allocations were more widely distributed throughout most of France. Only the Ile de France with its strong economy and rural Limousin were non-recipients. Through the *départements* (NUTS-3), 3.8 billion ECU was available to facilitate parts of regions affected by industrial decline. Objective 5b allocations were even more ubiquitous, vulnerable rural areas in need of economic diversification being allocated 2.2 billion ECU. Areas eligible for EU regional aid contained, in total, 47.6 per cent of the population of France, with 25.9 per cent living in Objective 2 areas, 17.3 per cent living in Objective 5b areas and only 4.4 per cent living in Objective 1 areas (CEC, 1996b).

Unitary states with planning power substantially devolved to the regions

Planning power has been substantially devolved to the regional tier of government in Italy and Spain – countries with a Napoleonic legal and administrative background. This has led since the Second World War to fragmentation and complexity in the case of Italy, while in Spain devolution has been the direct result of more recent pressures for constitutional reform, with a move to federalism and associated changes in planning powers and responsibilities. In both countries, but to a different extent, there is a combination of centralised control and a responsiveness to regional aspirations, which in planning has been reflected by 'the tendency to prepare a national code of . . . regulations and to create a hierarchy of plans based on a zoning approach' (Newman and Thornley, 1996: 72).

Italy

In the immediate post-war years, regionalism was promoted as a defence against the recurrence of fascism (Norton, 1983), although its development was slow since Christian Democratic governments feared that it would reduce their power at the centre (Wannop, 1995). Thus throughout most of Italy, Ministries of the central government held on tightly to power, or, in the case of the eight regions of the south (Campania, Abruzzi, Molise, Puglia, Basilicata, Calabria, Sicilia and Sardegna) established an agency, the *Cassa per il Mezzogiorno*, to help finance and develop irrigation, agriculture and industrial development in the most disadvantaged areas (Wannop, 1995). Nevertheless, regional economic committees were established in the mid-1960s, broadly, as in the UK and France, to provide a regional dimension to the national plan, and regional plans were produced by 1970. In the same year, elections took place in each of the regions leading to the formation of regional assemblies with a responsibility for planning and a range of other activities. Since 1972, therefore, although planning legislation is a responsibility of the national government, it has been reluctantly obliged to share this role with the regions. The national government, however, is principally concerned with strategic issues rather than with physical planning. It produces a series of sectoral plans, such as a transportation plan which provides guidelines and co-ordination for transport policies nation-wide, or development policies aimed at the South (Newman and Thornley, 1996). It also issues planning guidelines on other matters of national interest and major environmental impact, formulates policy on inter-regional balance, co-ordinates the administrative activity of the regions, and formulates planning standards (Ave, 1991).

Although the regions have gained legislative powers comparatively recently, they have fewer responsibilities than their counterparts in other regionalised states, for example, Belgium, Germany and Spain, and enjoy relatively little legislative and administrative autonomy (Wiehler and Stumm, 1995). Despite their legislative powers concerning railway net-works, roads of regional interest, regional communications, environmental and energy policy, public construction activity, housing and tourism, all matters of regional interest are subject to the national government's power to enact outline legislation and this limits the ability of the regions to pass their own legislation (Wiehler and Stumm, 1995).

Of the twenty regions (*regioni*), five, however, have special status (Friuli-Venezia, Sardegna, Sicilia, Trento-Alto Adige and Valle d'Aosta), with wider legislative and administrative powers over matters of regional con-cern, but in general the principal planning function of the regions is to regulate the system of physical planning applied by lower-tier authorities, the municipalities. Each region is therefore obliged to prepare a structure plan (*a Piani d'inquadramento* or *Piani-quadro*) which is intended to provide the guidelines for physical planning at the level of the municipality, but since the structure plan is very general and relates largely to economic and social issues it is often of limited use (Newman and Thornley, 1996).

The ninety-five provinces (*provinze*) have certain powers assigned to them by law, for example the preparation of structure plans (*Piani Territoriali di Coordinamento* under a new law primarily introduced to deal with local government boundary problems, the *Ordinamento delle Autonomie Locali* 1990. In contrast with the *Piani d'inquadramento*, this has provided a fairly effective framework for bringing together sectoral issues such as economic development and local planning issues. Metropolitan problems, however, were becoming a major cause of concern, therefore, under the 1990 Act, eleven *metropolitan cities* (with their own directly elected bodies) were given the same planning status as the provinces by becoming responsible for producing structure plans. This enabled the largest urban authorities to by-pass the regions and to implement policies directly in liaison with the national government (Mazza, 1991).

At the local level, the majority of the 8067 municipalities (communi) produce the principal planning instrument of the Italian planning system, the Master Plan (Piani Regolatore Generale). Dating back to the legal foundation for the Italian planning system in 1942, the Master Plan specifies land-use through a zoning process, defines principal communica-tion routes and identifies the location of new infrastructure. The specific

Figure 7.1 The regions of Italy

intentions of the Master Plan are subsequently implemented through a more detailed plan, the *piano particolareggiato*. Since the Master Plan is essentially concerned with physical and design issues, it is the key instrument in setting out objectives and guidelines concerning new development and construction throughout the commune area (Newman and Thornley, 1996), but because there is no time limit on the plan, changes required subsequently need to be considered case by case, often in a haphazard manner.

Since some municipalities are too small to deal with planning matters, their responsibilities have been subsumed by the provinces, while the metropolitan cities have become quasi provinces with planning powers of their own. Although this might indicate that there is a partial shift from a fragmented approach to planning to a more comprehensive framework, north-south divisions are increasingly being emphasised by political parties such as the neo-fascist MSI or the Northern League to increase the degree of fragmentation or to break up the Italian state, the latter party proposing initially to reassemble the twenty Italian regions into only three (Wannop, 1995), but subsequently urging the formation of an independent state, 'Padania', comprising the regions of Piemonte, Lombardia, Trentino-Alto Adige, Friuli-Venezia Giulia, Liguria, Emilia Romagna, Veneto, Toscana, Umbria and Marche.

Clearly, without a common ethnic, linguistic or cultural heritage, Italy continues to be essentially a *regionalised state* (Wiehler and Stumm, 1995) and the intra-regional fragmentation of planning remains a practical reality. With regard to enhancing the role of regional government, there is considerable scope for reform, short of the political disintegration of Italy.

Regional policy and regional incentives

For several decades, regional policy in Italy concentrated on regenerating the geographical south through the medium of the *Casa per il Mezzogiorno*. The Casa had powers to initiate development projects and to co-ordinate development in every sector of the southern economy. In addition to funds being allocated by the central government, investment was also forthcoming from the private sector, both Italian and external, and by loans from the World Bank and, since the formation of the EEC, from the European Investment Bank, the Agricultural Guidance Fund and the Regional and Social Funds (Minshull, 1996). By the time of its liquidation in 1985, if not before, it was clear that the Casa had not succeeded in narrowing the economic disparities between the north and south. Manufacturing industry had not been developed in the south on the scale required and tended to be capital-intensive rather than labour-intensive, with an emphasis on public utilities (including power, water and oil-refining), rather than on modern private-sector industries with the ability to develop linkages with other industries, such as engineering or chemicals (Hall, 1992). Low productivity remained in agriculture, and unemployment stayed at a level considerably higher than the national average. With little doubt, 'a stronger policy, with some negative controls on industrial growth in the more prosperous regions closer to the heart of the (then) EEC, would seem to [have been] . . . necessary' (Hall, 1992: 189).

After the demise of the *Casa*, its powers were transferred by new legislation to regional governments in 1986. The Mezzogiorno was divided into three grades of area, and eligibility for aid was differentially extended through the medium of three-year rolling programmes. There was an attempt to promote small-scale industry and tourism, facilitate irrigated agriculture, and rationalise development in mountain areas, in each case with the aid of EU funding (Minshull, 1994). In 1990, under pressure from the EC, parts of the North Mezzogiorno were de-designated, and others from 1992, although discriminatory rates of support were increased in the remaining areas. Finally, 'special intervention' for the Mezzogiorno was abolished in 1993 (CEC, 1994c).

Despite the application of regional aid over a period of more than forty years, the south, in the early 1990s, still remained an economically disadvantaged region when compared to Italy and the (then) EC as a whole. Table 7.1 shows that the eight southern regions had a disproportionately large percentage of employment in agriculture, and a less than proportionate amount of employment in industry, reflecting the imbalance in development described above. Table 7.1 also reveals that the south had disproportionately high rates of unemployment, and that the level of unemployment had, in fact, increased over the period 1986–93 in seven of the eight regions reflecting the fragility of the southern economy, and Table 7.2 indicates how the poorly performing economy of the south is reflected in low per capita GDPs.

Clearly, when the distribution of EU Structural Funds for the period 1994–99 was decided, it was southern Italy which qualified for the whole of the country's Objective 1 allocation (see Figure 7.2). The development of the eight regions was undoubtedly lagging behind. Through the regional tier of government (NUTS-2), a sum of 14.9 billion ECU was available to promote development and structural adjustment, although in the case of Abruzzi funding ceased in 1996 when the region reverted to Objective 2

Table 7.1 Employment and unemployment in the southern regions of Italy

	Employment (%) 1990			*Unemployment (%)*		
	Agriculture	*Industry*	*Services*	*1986*	*1993*	*Change 1986–93*
Campania	15.2	21.6	63.3	16.6	22.8	+6.2
Abruzzi	14.6	26.6	58.8	11.7	12.4	+0.7
Molise	20.9	26.1	53.0	7.1	15.6	+8.5
Puglia	12.8	24.0	63.2	14.3	15.6	+1.3
Basilicata	22.6	25.6	51.8	21.0	23.0	+2.0
Calabria	25.7	18.0	56.3	15.4	19.6	+4.2
Sicilia	14.6	19.8	65.7	15.1	23.1	+8.0
Sardegna	13.7	22.1	64.2	20.2	19.8	−0.5
Italy	9.6	29.7	60.7	10.5	11.2	+0.6
EC 12	6.6	32.1	61.3	10.7	10.4	−0.3

Source: CEC (1994)

Table 7.2 Gross domestic product per capita in the southern regions of Italy

| | *(EC 12 = 100)* | |
	1986	*1991*
Campania	69	70
Abruzzi	89	91
Molise	78	79
Puglia	73	74
Basilicata	65	65
Calabria	60	57
Sicilia	70	68
Sardegna	75	74
Italy	103	105
EC 12	100	100

Source: CEC (1994)

status. Investment was targeted at eight priorities reflecting acknowledged deficiencies:

1   Industry, small medium enterprises, craft businesses, business services and industrial estate development (25 per cent).
2   Infrastructure to support economic activity, water resources, energy networks, renewable energy, water and waste management, research and development (22.2 per cent).
3   Diversification and development of agricultural resources and rural development (15.7 per cent).
4   Communications: road and rail improvements linked into trans-European networks, and telecommunications (14.5 per cent).
5   Development of human resources through education, programmes for the long-term unemployed and training for public administrators (14.5 per cent).
6   Tourist development and the natural and cultural environment (5.8 per cent).
7   Fisheries (1.7 per cent).
8   Technical assistance, publicity and monitoring (0.6 per cent).

Operational Programmes for each priority and for each region were produced to ensure that the large volume of Objective 1 funding was effectively targeted and managed over the period of allocation. By comparison, the amount of Objective 2 and 5b funding was small. Objective 2 areas, the provinces (NUTS-3), were eligible for assistance of up to a total 1.5 billion ECU to convert local economies seriously affected by industrial decline, and Objective 5b areas were eligible for only 901 million ECU to facilitate the development and structural adjustment of rural areas. Applicant regions and

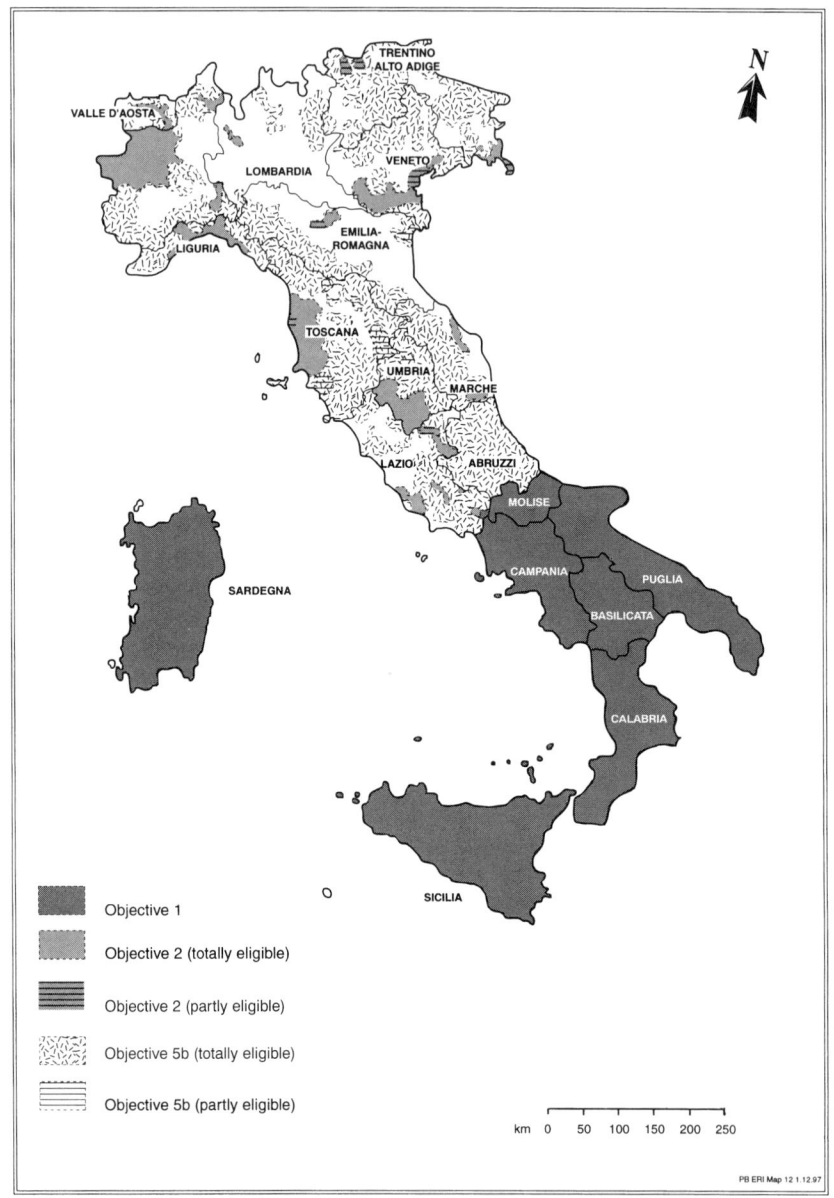

Figure 7.2 Designated Structural Fund areas, Italy, 1994–99

the provinces were obliged to submit development and conversion plans to the central government for consideration prior to seeking funding from the Commission.

At the time of designation, Objective 1 regions contained 36.6 per cent of the Italian population, Objective 2 provinces contained 10.2 per cent and only 8.4 per cent lived in Objective 5b areas. In total, areas benefiting from EU regional funding contained 55.8 per cent of the Italian population (CEC, 1996b).

Spain

Despite the unification of Spain under the dominance of Madrid and Castile in the sixteenth century, strong regional cultures remained around the periphery, but were frequently suppressed, most notably in the period 1939–75 under the Franco regime. Nevertheless, throughout most of the twentieth century, fifty provincial administrations and seven island councils provided the middle tier of government, between the central state and the country's several thousand municipalities. Until 1925, provincial adminis- trations acted as agents of the central state, akin to the French model, but thereafter became partners rather than controllers of municipal government (Wannop, 1995).

During the latter years of the highly-centralised Franco regime, the planning system of Spain was under increased pressure from large-scale migration to urban areas, corruption, illegal development, local authority negligence and chaotic urban growth (Wynn, 1984; Newman and Thornley, 1996). Clearly the system was under stress, and its inability to cope with urban growth was compounded by the unwillingness of the municipalities to co-operate with their neighbours. Very few boards were established to counter the problems of municipal fragmentation, although the central state, in 1963, transferred the responsibility of planning the expanding conurba- tion of Madrid from the municipalities to a newly established central agency, COPLACO, rather than creating a metropolitan government (Wannop, 1995). In Barcelona, in contrast, a metropolitan multi-purpose authority was established for an area containing twenty-seven municipalities and a population of two million. With the termination of long-term macro- economic planning in 1973, it was considered necessary to offset the effects of a freer market economy by creating greater certainty in the future use of land (Wannop, 1995). At both regional and provincial level, statutory regional plans were therefore introduced replacing the earlier unsuccessful advisory plans (Teixidor and Hebbert, 1982). The planning system of Spain had also been unsuccessful in steering investment and new jobs to disadvan- taged areas, notably in the rural periphery, despite functional regionalism and centralised incentives (Clout *et al.*, 1994).

In an attempt to combat the inequalities and inefficiencies associated with uncontrolled urban growth and regional imbalance, the principal instrument of planning policy, the ineffective planning Act of 1956, was finally replaced by an amended Act in 1975, but with the introduction of a new Constitution in 1978, there was an opportunity to introduce a more effective and devolved system of planning. Although the national government assumed responsibility for such matters as major infrastructure development, national parks and for formulating guideline legislation for a wide range of policies, a generally high degree of autonomy was granted to the seventeen autonomous regions of Spain (*comunidades autonomas*), the development of regional administrations, as in Italy, being seen as a defence against the return of fascism (Norton, 1983). Each region in Spain has its own elected government (*a junta*) and legislative assembly; has responsibilities (sometimes shared with national government) for regional economic and urban development, planning, infrastructure and environmental protection, is consulted when the national plan is being established, and implements policies in accordance with its Regional Economic Programme and the Guidelines on the Development of Territory. The extent of autonomy, however, varies from region to region and is greatest in the four autonomous regions which enjoy wider legislative powers (Andulucia, the Basque region, Galicia and Cataluña) (Wiehler and Stumm, 1995). By the 1990s, however, the scope of the *juntas* to formulate planning policy was increasingly being questioned by national government, and was under review by the Constitutional Court (Keyes *et al.*, 1991; Newman and Thornley, 1996). With over a dozen regional plans being produced, there was a risk that they would be unco-ordinated and, in consequence, detrimental to the national interest (Clout *et al.*, 1994).

Sandwiched between the autonomous regions and the municipalities, a layer of provincial administrations attempted to maintain their relationship with the municipalities which they had hitherto possessed, but since the *juntas* resourced local government in conjunction with central government and had powers to reform it, some wished either to amalgamate the provinces into larger and more efficient units or to minimise their role completely (Wannop, 1995). In Andulusia, for example, the number of provinces has been reduced to only nine (it was previously more than twice that number), while in Cataluña the multi-purpose metropolitan authority for Barcelona was sub-divided functionally, a different metropolitan council being responsible for each of the principle local services. In contrast, other *juntas* increased the importance of the provincial tier of government, for example in the region of Valencia, its *junta* established a multi-purpose province for the metropolitan of Valencia, but in the national capital, the central government relinquished its administrative responsibility for the city by permitting the *junta* of the region of Madrid to assume the former responsibilities of the government's agency, COPLACO (Wannop, 1995).

Under the 1992 Act (the *Texto Refundido de la Ley Sobre el Régimen de Suelo y la Ordencion Urbana*), planning at a local level became the responsibility of the 8077 municipalities (*municipios*). If their populations exceed 5,000, the

Figure 7.3 The regions and provinces of Spain

municipalities are obliged to produce legally binding general urban plans showing land zoned into three categories: land excluded from development, land available for development and land already developed. The middle category is sub-divided into programmed land (land which is required to be developed in accordance with the general plan) and non-programmed land (land for which no programme or strategy has been adopted for it and which might be developed after the general plan has been implemented). In addition, the general plan describes the specific use of each zone (low/ medium or high rise residential, heavy and light industry etc.), designates transport networks, and indicates the location and siting of public buildings, conservation areas and open space. Plans for both programmed and non-programmed land need to be in two stages of four and eight years (Bassols, 1986; Newman and Thornley, 1996).

Although general urban plans have to be approved by the regional authority to ensure that they conform to higher order plans and regulations, at a very local level the municipalities are able to control development without higher level approval. The *plan parcial*, for example, which has to be produced before programmed urbanised land is developed, is not subject to approval by the region, neither are other instruments such as analytical memoranda that accompany plans, or detailed studies which supplement the general plan or the *plan parcial*.

Since 1978, development control has been much tighter than hitherto. Licences are required for all development, demolition and land sub-division, and applicants need to satisfy the issuing municipality that their proposals conform with both plans and building regulations, a task which might need support from the college of architects if a major development is being considered (Newman and Thornley, 1996). In respect of non-programmed land, a more flexible approach is adopted, since a developer, in seeking approval for a proposal, need not conform to specific land-use requirements.

Undoubtedly, the 'complex system of laws and instruments in Spain provides a degree of certainty, particularly at the municipal level', and that, with the aim of reducing regional disparities, 'national and regional plans for major infrastructure projects are . . . contributing to the develop-ment of transitional networks for transport, energy and telecommunication' (European Commission, 1994). Notwithstanding these benefits, and con-trary to the requirements of planning legislation, it is far from certain that a hierarchy of plans will emerge in which lower level plans conform to the higher ones. In a climate of regional autonomy, it is unlikely that a national plan will emerge, while at the same time regional plans have been slow to materialise leaving the municipalities with little higher-level guidance (Newman and Thornley, 1996; Keyes *et al.*, 1993).

Regional policy and regional incentives

Like Italy, regional economic disparities in Spain have been very evident for generations. Currently, the contrasts are between the northern coastal area comprising the regions of Pais Vasco (the Basque Country), Cantabria and Asturias – areas not particularly poor but suffering from de-industrialisation, Andulucia in southern Spain – a region lagging behind in terms of economic growth, and the Mediterranean coast comprising the regions of Cataluña, Comunidad Valenciana and Murcia – an area experiencing varying degrees of economic growth (Drake, 1994).

Since the 1980s, regional policy in Spain has aimed to facilitate the development of the economically disadvantaged areas, particularly in the northern coastal area and in Andulucia. In 1986, legislation introduced new regional development incentives and specified problem regions eligible for assistance, and in 1987 new zones eligible for regional assistance were specified and maximum rates of support were set for each grade of zone – ranging from 20 to 50 per cent (or 75 per cent in Zones of Industrial Decline). In 1991, the spatial allocation of aid was rationalised by the incorporation of the Zones of Industrial Decline into the more general Zones of Economic Promotion (CEC, 1994c).

Despite the development of regional policy over the previous ten years, by the early 1990s it was evident that as many as ten regions remained severely disadvantaged in terms of employment distribution, unemployment and per capita GDP in comparison with both Spain as a whole and the twelve members of the (then) EC. Table 7.3 reveals that agricultural employment in these regions was often disproportionately high (reflecting the sector's high labour intensity), while industrial employment was often disproportionately low (suggesting industrial underdevelopment). Table 7.3 also shows that unemployment was high and had increased in almost all regions over the period 1986–93, and Table 7.4 indicates the extent to which the GDP per capita in these regions was lagging behind national and EC levels.

Since economic development in the ten autonomous regions (NUTS-2) listed in Tables 7.3 and 7.4 was clearly lagging behind, each became eligible for Objective 1 funding for the period 1994–99 (see Figure 7.4).

Table 7.3 Employment and unemployment in the disadvantaged regions of Spain

	Employment (%) 1990			Unemployment (%)		
	Agricultural	Industrial	Service	1986	1993	Change 1986–93
Gallicia	31.5	23.5	45.0	13.9	17.0	+3.1
Asturias	15.2	32.5	52.3	18.8	19.6	+0.9
Castilla-León	18.9	28.8	52.3	18.1	19.2	+1.0
Castilla-La Mancha	18.5	31.9	49.6	15.4	18.6	+3.2
Extremadura	24.3	21.3	54.5	28.3	28.9	+0.7
Comunidad Valenciana	8.7	36.8	54.5	19.7	22.8	+3.1
Andalucia	14.8	25.2	60.0	30.3	30.8	+0.5
Murcia	13.1	31.0	55.9	18.4	23.4	+4.9
Ceuta y Melilla	0.8	9.6	89.5	28.7	21.9	−6.8
Canarias	8.0	20.8	71.2	26.5	26.7	+0.1
Spain	11.0	32.1	56.9	21.4	21.3	−0.1
EC 12	6.6	32.1	61.3	10.7	10.4	−0.3

Source: CEC (1994)

Table 7.4 Gross domestic product per capita in the disadvantaged regions of Spain

	(EC12 = 100)	
	1986	1991
Gallicia	56	59
Asturias	71	73
Castilla-León	66	67
Castilla-La Mancha	55	64
Extremadura	45	51
Comunidad Valenciana	72	78
Andalucia	53	60
Murcia	68	74
Ceuta Y Melilla	64	64
Canarias	70	77
Spain	71	78
EC 12	100	100

Source: CEC (1994)

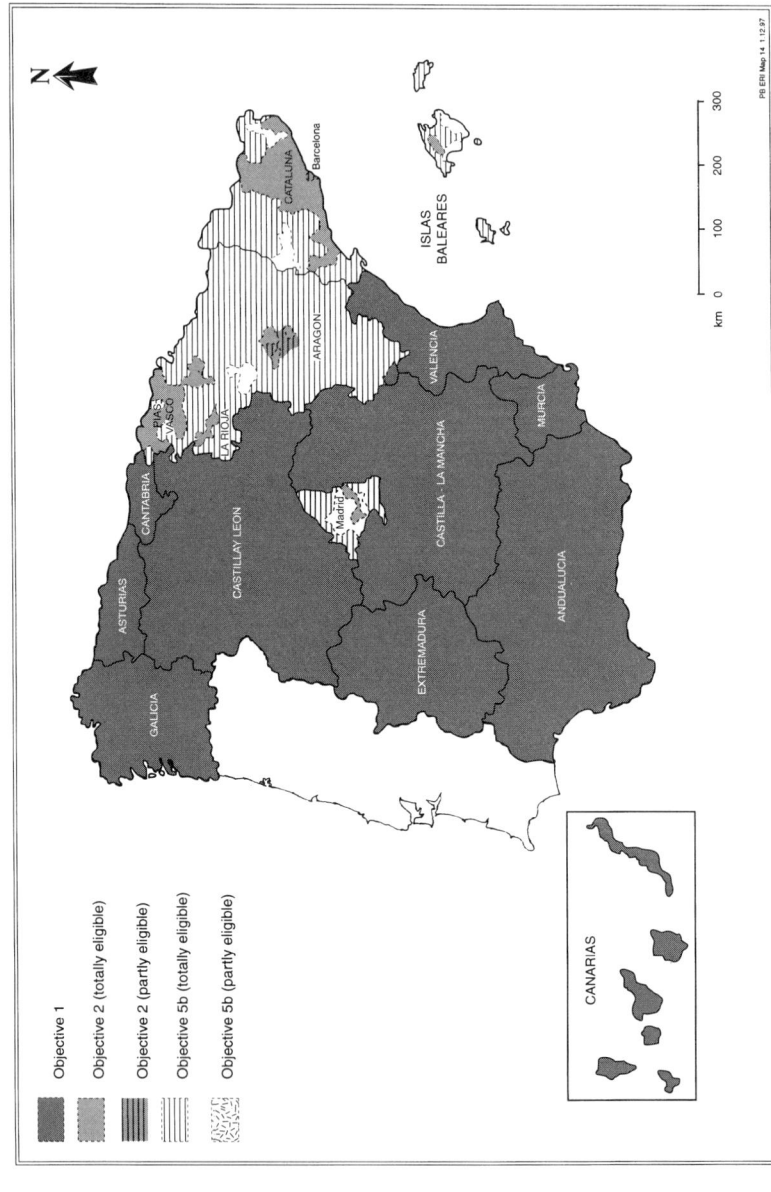

Legend:

Objective 1
Objective 2 (totally eligible)
Objective 2 (partly eligible)
Objective 5b (totally eligible)
Objective 5b (partly eligible)

GALICIA
ASTURIAS
CANTABRIA
PAIS VASCO
LA RIOJA
CASTILLA Y LEON
CATALUNA
Barcelona
ARAGON
MADRID
Madrid
EXTREMADURA
CASTILLA - LA MANCHA
VALENCIA
MURCIA
ANDALUCIA
ISLAS BALEARES
CANARIAS

km 0   100   200   300

PB ERI Map 14 1.12.97

Figure 7.4 Designated Structural Fund Areas, Spain, 1994–99

A total of 26.3 billion ECU was available to facilitate development and structural adjustment. To target aid at areas of particular concern, four priorities were set:

1 To improve the production system (34.5 per cent of funding).
2 To develop human resources through education, training and health services (33.4 per cent of funding).
3 To reduce the isolation of remote regions through the development of the road network, improvement to the rail system, port and seaport development and telecommunications (24.8 per cent of funding).
4 To develop energy networks and water management (7.3 per cent of funding).

A total of fourteen Operational Programmes were subsequently produced to facilitate and manage development during the period of funding. By contrast, eligible provinces (NUTS-3) received only a limited amount of Objective 2 funding, 2.4 billion ECU for the period 1994–99, to facilitate the economic conversion of areas of industrial decline, while Objective 5 areas were eligible for the least funding, only 664 million ECU, to facilitate rural development and structural adjustment. Applicant regions and provinces were obliged to submit development and conversion plans to the central government for consideration as a preliminary to seeking funding from the Commission.

Areas eligible for regional aid contained, in total, as much as 82.9 per cent of the population of Spain (an even higher proportion than Italy), with 58.2 per cent living in Objective 1 areas, but as little as 20.3 and 4.4 per cent living respectively in Objective 2 and Objective 5b areas (CEC, 1996b).

8 Federal states with planning power largely vested in the regions

Within the Federal states of Austria, Switzerland and Germany planning power, to a considerable extent, is vested in the regional tier of government. Although there is a marked variation in the planning processes between the three countries, the strong constitution and Federal system of each member of the Germanic family is similarly legalistic in its approach – the codification of law being reflected in 'rigorously formulated planning regulations . . . (and) . . . a strong level of planning with its own laws and plans and a set of arrangements for creating consensus between and within levels in the hierarchy' (Newman and Thornley. 1996: 72). As in Italy and Spain, the devolution of planning power to the regions in the *Napoleonic* state of Belgium has been the direct result of recent pressures for constitutional reform and associated changes in planning powers and responsibilities – culminating in the creation of a Federal state in 1993.

Austria

The three-tier Federal administrative structure of Austria is broadly similar to that of Germany. At the upper level of the three-tier administrative structure of Austria, the Federal state exercises control over matters of national importance, for example, finance, foreign policy, defence, education, the national road system, railways and flood control.

The middle tier consists of nine regions *(Länder)*, all of which were established after the First World War as administrative entities. Although each region has its own Parliament and government, the regions in total accounted for only about 27 per cent of government expenditure in the mid-1990s and their powers are comparatively weak. The middle tier of government, from its formation, was undoubtedly handicapped by the suppression of regional identity and political representation during the period of the Austro-Hungarian Empire prior to 1919. The regions, nevertheless, became responsible for collecting planning information, preparing regional plans and ensuring that plans are co-ordinated with the aims and objectives of the central government and neighbouring regions. Regional plans can either cover the whole of the area of the *Länder* or there may be several sub-regional plans as well as subject plans (Newman and Thornley, 1996). Although each region implements its plans in a different way, planning is in accordance with the same general legal restrictions.

At the lower-tier, or local government level, Austria is divided into 84 regional districts (*Landbezirke*) and fifteen urban authorities, below which there are 2350 municipalities. Although the regional districts and urban authorities might be subject to sub-regional plans, the municipalities have the responsibility of producing local physical plans setting out the rationale of development, means of implementation and a blueprint for the use of all land within the municipality – each plot being subject initially to a general land-use classification such as building land or green belt land before being sub-divided into more specific uses. Providing that the municipal plan conforms to the regional plan and is financially sound, it is subsequently approved by the region and becomes legally binding on individuals (Newman and Thornley, 1996).

Since the Federal, *Länder*, and local governments are all involved in regional policy, co-ordination is clearly necessary – a task performed by the Austrian Conference on Regional Planning (ÖROK).

Regional policy and regional incentives

In recent years, regional policy objectives in Austria have been more varied than in the larger nations of Western Europe, and have tended to change as the international context has altered and the economy has developed (CEC, 1994). In the absence of major disparities in, for example, unemployment and GDP per capita, the aims of regional planning in Austria include the protection of the environment, most notably in the Alpine *Länder,* the re-orientation of international transport networks, particularly rail links to Central and Eastern Europe, and cross-frontier co-operation in the border regions, especially where there are agriculture and declining industries.

Only one *land* was deemed to be lagging behind the rest of the economy – Burgenland, in the extreme south-east adjacent to the Austro-Hungarian border, a region suffering from a disproportionate reliance on agriculture (see Figure 8.1). As a NUTS-2, Burgenland became eligible for Objective 1 funding (with a sum of 162 million ECU being available, 1994–99). In addition, a number of *Landbezirke* (NUTS-3) in Syria, Lower Austria and Upper Austria became eligible for Objective 2 funds amounting to 99 million ECU to facilitate the economic conversion of areas of industrial decline during the period 1994–99, but vulnerable rural areas attracted most of the Structure Funds allocated to Austria, Objective 5 areas being eligible for 403 million ECU to diversify economic activity to the end of the century. Applicant *Länder and Landbezirke* were obliged to submit development and conversion plans to the Federal government prior to seeking funding from the Commission.

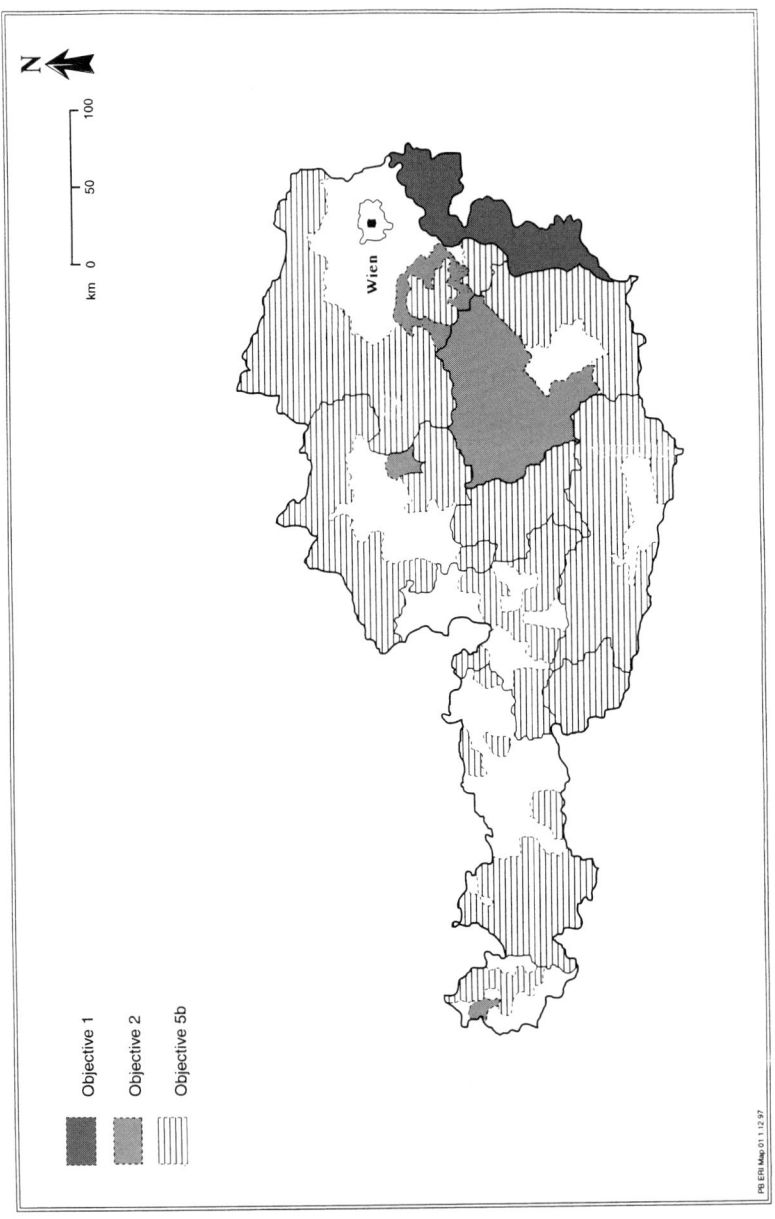

Objective 1

Objective 2

Objective 5b

km 0 50 100

Wien

N

PB ERI Map 01 1 12 97

Figure 8.1 Designated Structural Fund Areas, Austria, 1994–99

In total, areas eligible for regional funding contained, in 1994, 40.6 per cent of the population of Austria with Objective 5b areas containing 28.9 per cent, Objective 2 areas containing 8.2 per cent and the sole Objective I area containing only 3.5 per cent (CEC, 1996b).

Switzerland

Switzerland is also a Federal state, divided into three tiers of government, each of which is involved in planning. A general law on the aims and objectives of planning at the national or confederation level was incorporated into the Federal Law on Spatial Planning 1980, while national government was also required to co-ordinate its own activities with the planning responsibilities of the middle tier of government (Newman and Thornley, 1996).

Co-ordination, however, was not very successful (Ringli, 1992, 1994). As a consequence, in 1995, the national government introduced a strategy called *Guidelines for Swiss Spatial Development* to provide a framework for planning at the middle level of government. The strategy was designed to enable Swiss cities to compete successfully with comparable urban areas outside the borders of Switzerland, such as Frankfurt, Munich and Lyons. More specifically, the strategy aimed to create an integrated *Swiss City* of some 3 million inhabitants comprising most of the cities and towns of Switzerland tied together by an improved public transport network. New commercial development, it was hoped, would become concentrated on public transport nodes, while restrictions on the peripheral growth of each city would prevent congestion and enhance the attractiveness of the countryside (Newman and Thornley, 1996). A highly efficient and environmentally attractive urban structure would thus emerge and attract top management and senior professionals and their respective families (Ringli, 1994).

Twenty-six cantons comprise the middle or regional tier of government – the Helvetic Confederation being originally formed by the cantons of Schwyz, Uri and Unterwalden (Niedwalden and Oberwalden) in 1291. The last canton was founded by dividing the canton of Bern into two parts and form' the canton of Jura in 1978 (Lawrence, 1996). The cantons, each with a considerable degree of governmental independence, formulate their own Building and Planning Laws and are responsible for producing regional or guiding plans. As the principal instrument of spatial co-ordination within the confederation, regional plans, however, need to be approved by the Federal Council before becoming binding on authorities at all levels (Newton and Thornley, 1996).

At the lower level of government, some 3072 communes are responsible for controlling development by means of building regulations and zoning plans. Zoning plans, which have to be approved by the canton, specify the permitted use of all land within the commune and are legally binding on all landowners.

Had Switzerland voted in favour of joining the EU in recent years, very few areas would have been eligible for Structural Funds since there is an absence, within the Federal state, of regions lagging behind or areas in need of economic conversion due to industrial decline. It is possible, however, that some of the Alpine and Jura areas would have been eligible for Objective 5b funding to facilitate development or conversion within the rural economy.

Germany

Regional planning in Germany has a long history dating back to the early years of the twentieth century. In 1910, a common purpose planning authority for Greater Berlin (the *Zweckverband Gross-Berlin*) was established to plan the future development of the city's metropolitan region which already had a population of 3.5 million, and in 1920 the spatial extent of the planning authorities' responsibilities was extended as Greater Berlin became the second largest city in the world by the absorption of 93 suburbs (Wannop, 1995). But it was in the west of Germany that regional planning made the most progress. Founded in 1920 in an area containing a population of 10 million, the Ruhr Coalfield Settlement Association (the *Siedlungsverband Ruhrkohlenbezirk),* a non-elected but representative authority, promoted regional development, undertook transport planning and environmental improvement, applied a recreational strategy and co-operated with urban authorities in the production of town plans (Wannop, 1995). There was an absence, however, of a nationwide system of regional planning throughout the period of the Wiemar Republic (1919–33), and during the subsequent Third Reich the centralised state became paramount.

The contemporary spatial planning system of the Federal Republic of Germany, founded after 1945, is a reflection of Germany's exceptionally decentralised administrative system which was deliberately created as a reaction to the centralised power of the former state. The Federal planning system operates through the medium of a decentralised decision-making structure and a robust legal framework with a strong emphasis on codification and the interpretation of planning law (Hooper, 1989; Newman and Thornley, 1996). Like most other functions of government in the republic, spatial planning is administered at three levels stemming from the constitutional division of power set out in the Basic Law *(Grundgesetz)* of 1949, although this often necessitates the upper tier of government having to advise the lower tiers on the intricacies of the planning system.

Notwithstanding the need for subsidiarity (which enables lower-tier authorities to exercise a degree of autonomy in relation to the details of policy), it is a basic requirement of the system that the policies and plans of lower-tier authorities have to conform to those of the higher tier (Newman and Thornley, 1996).

The Federal Ministry of Region Planning, Building and Urban Development (BMBau in its German abbreviation) and representatives of *Land* governments annually draw up a general document of national guidelines and co-ordinate the specialised planning responsibilities undertaken at the Federal level and the planning measures applied by the *Länder* as middle tier authorities. Framework legislation is consequently introduced in the form of

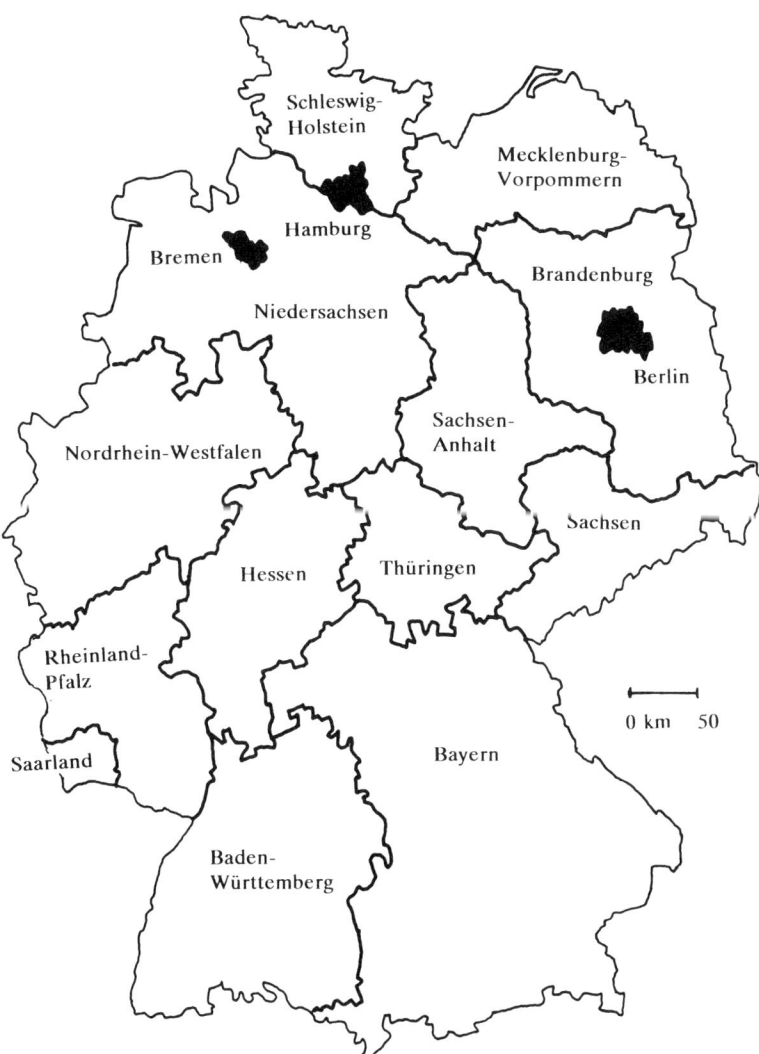

Figure 8.2 The *Länder* of Germany

Federal Regional Planning acts *(Bundesraumordnungsgesetz)* for all of the *Länder,* and by means of the Federal Building Code *(Baugesetzbuch)* of 1986 (amended in 1993) which regulates development projects and building proposals – 'the most important piece of legislation in relation to urban development in Germany' (EC, 1994: 148).

Notwithstanding considerable variation planning practice and planning law at the middle level of government (despite Federal attempts to establish consistency), each of the sixteen *Länder* is obliged to produce a *Landesplannungsgesetz* (a regional planning act) and prepare and adopt a *Landesentwicklungsprogramm* (a regional comprehensive development programme). Derived from this, each *Land* prepares a *Landesentwicklungspläne* (a regional comprehensive development plan) and a regional *pläne,* and where appropriate a number of comprehensive sub-regional plans.

Comprehensive development plans normally include broad statements concerning population projections, settlement hierarchies, priority areas and intended development, whilst more detailed statements are included in sub-regional plans.

At the local government level, the 552 counties *(Kreise)* have few planning powers and only undertake some of the broader functions of lower-tier government which the smaller authorities cannot fund, such as major roads and hospitals (Newman and Thornley, 1996). The 16,068 towns and municipalities *(Städt* and *Gemeinden),* however, produce two types of plans, a *Flächennutzungsplan* (a preparatory land-use plan) – binding on the public authority but with no legal affects on the rights of individuals, and a conforming *Bebauungsplan* (a legally binding land-use plan) which, following consultation with local people and with the relevant *Land* to ensure conformity with regional objectives, can control, for example, the amount of developable floorspace, methods of construction, materials, the height of buildings, infrastructure and car-parking. Building and planning control is dependent upon the issue of a single permit granted by the local authority, subject to the applicant complying with the relevant planning regulations (EC, 1994), and although a right of appeal exists against the refusal of a permit, such an appeal must be based on a legal irregularity rather than on the planning merits of the proposed development.

An overarching objective of the Federal spatial planning system is to create 'equivalent living conditions throughout the Republic' (EC, 1994: 148), an objective facilitated by a financial equalisation system which provides for a transfer of funds from high per capita income to low per capita income *Länder.*

This objective has become even more important since the reunification of Germany on 3 October 1990. The five *Länder* and 7564 *Städt* and *Gemeinden* of the former German Democratic Republic (GDR) were absorbed into the Federal Republic, with the result that severe disparities between the western and eastern regions of the new Germany became apparent, particularly in terms of economic growth and employment levels. Disparities have been compounded by migration from East to West – undermining economic restructuring in the East and exacerbating housing shortages in the West (EC, 1994). In response to political pressure to transform the centrally planned economy of the former GDR into the market economy of the European Union, the *Spatial Planning Concept for the Development of the new Länder* (BMBau, 1992, English edition) indicated that, in order to avoid delays, the Federal government intended to adopt a more direct and rapid procedure than that utilised in the western *Länder*. To help narrow the East-West divide and to create a spatially balanced settlement structure, a spatial development strategy was introduced to provide a framework for the governments of each of the new *Länder* (Williams, 1996). The aim was to develop several major urban centres as counter-magnets to Berlin, involving the relocation of Federal and other public sector offices, universities and research establishments to the different urban centres, together with policies for rural development (Williams, 1996).

Encouraged by the experience of applying the spatial planning concept to the new *Länder,* the Federal government issued the *Guidelines for Regional Planning* (BMBau, 1993, English edition). Although intended as the basis for Federal spatial planning, the guidelines are very much influenced by the measures adopted to deal with the problems of reunification and, in large part, were written with regard to the wider European context particularly in respect of trans-border co-operation – not only within the EU but with Switzerland, the Czech Republic and Poland (Williams, 1996).

Regional policy and regional incentives

For the last fifty years or more, the spatial economy of Germany has been broadly divided into three main parts. First, the Ruhr coalfield of Nordrhein-Westfalen which has suffered badly as a result of the decline of the coal and steel industries; second, the growth areas of the Bayern and Baden Württemberg in the southern half of the country; and third, the eastern *Länder*, economically disadvantaged both before and after German reunification in 1990. Although there are other areas of comparatively fast growth such as Hamburg and Hessen, and declining areas such as Rheinland Pfalz/Saarland in the west, the eastern and Alpine borderlands and the coastal areas of West Germany, essentially there is (unlike Britain) both a north-south divide, and a west-east divide.

By the 1970s, three types of areas were qualifying for the largest share of regional aid development areas *(Bundesbaugebeite)*, development centres *(Bundesbauorte)* and the frontier zone with the (then) DDR *(Zonenrandgebiet)*. In addition, help was also given to the peripheral areas of Emsland in the north-west, the North Sea coast, northern Schelswig-Holstein and the Alps in the south. Focusing on these areas, the Federal government invested heavily in infrastructure, particularly in improved communications, and provided financial incentives such as investment allowances and grants to encourage private investment in industry. Region aid emanates initially from the Federal government, but is then administered through the *Länder* authorities, although in the case of the structural conversion of the Ruhr coalfield, the European Coal and Steel Community was responsible for distributing re-adaptation grants to redundant miners (Hall, 1992). In contrast, within the coalfield, the *Land* of Nordrheln-Westfalen, through the medium of the Emscher Park Planning Company, has allocated grants to organisations since 1988 to enable them to undertake a wide range of development projects in accordance in co-operation with local municipalities *(Gemeinden)* (Drake, 1994).

By the 1980s, it was apparent that regional policy had been generally successful, therefore it was inevitable that Federal aid would be gradually reduced and redirected. In 1981, therefore, the assisted areas were reduced from 60 to 36 per cent of the area of the Federal Republic, and their population decreased from 49 to 30 per cent of the country's total (Hall, 1992), while special investment grants were introduced to encourage the development of high-grade jobs. In 1982, a special steel location programme was introduced for a three-year period, and in 1984, under Commission pressure, assisted areas with less than 1.5 per cent of the total population were de-designated, while Bremen and Gelsenkirchen were designated as high incentive areas due to their respective industrial problems. In 1985, measures were introduced to promote service-related activities, innovation and research and development with an emphasis on infrastructure, and in 1988 a special medium-term programme for the mining regions was introduced. Also in 1988, however, aid ceilings were reduced, and investment allowances were abolished in 1989 although there was an increased budget for investment grants (CEC, 1994c).

With the reunification of Germany, there was a shift of emphasis in regional policy from assisting areas of economic disadvantage in the west to regenerating the economy of the former DDR, with aid being allocated in decreasing proportions to 'problem areas', A, B and C (see Figure 8.3). The differences between the two parts of Germany were substantial in terms of standards of living, industrial productivity, infrastructure and environmental quality. Because of over-manning, out-dated plant and machinery, poor problems and lack of marketing experience, it was soon evident that the industries of the east were unable to compete with those of the west (Drake, 1994). Output consequently fell dramatically and registered unemployment soared to 15 per cent by 1993 (well above the average rate in Germany), while large-scale migration took place to the west. The five eastern *Länder* and Berlin were consequently designated as assisted areas by the Federal government and became eligible for 21 billion ECU of investment grants – a minute sum compared to an estimated £500 billion of tax revenue that will need to be transferred from the west to the east of Germany by the year 2000 (Drake, 1994).

When the Commission allocated Structural Funds to member countries

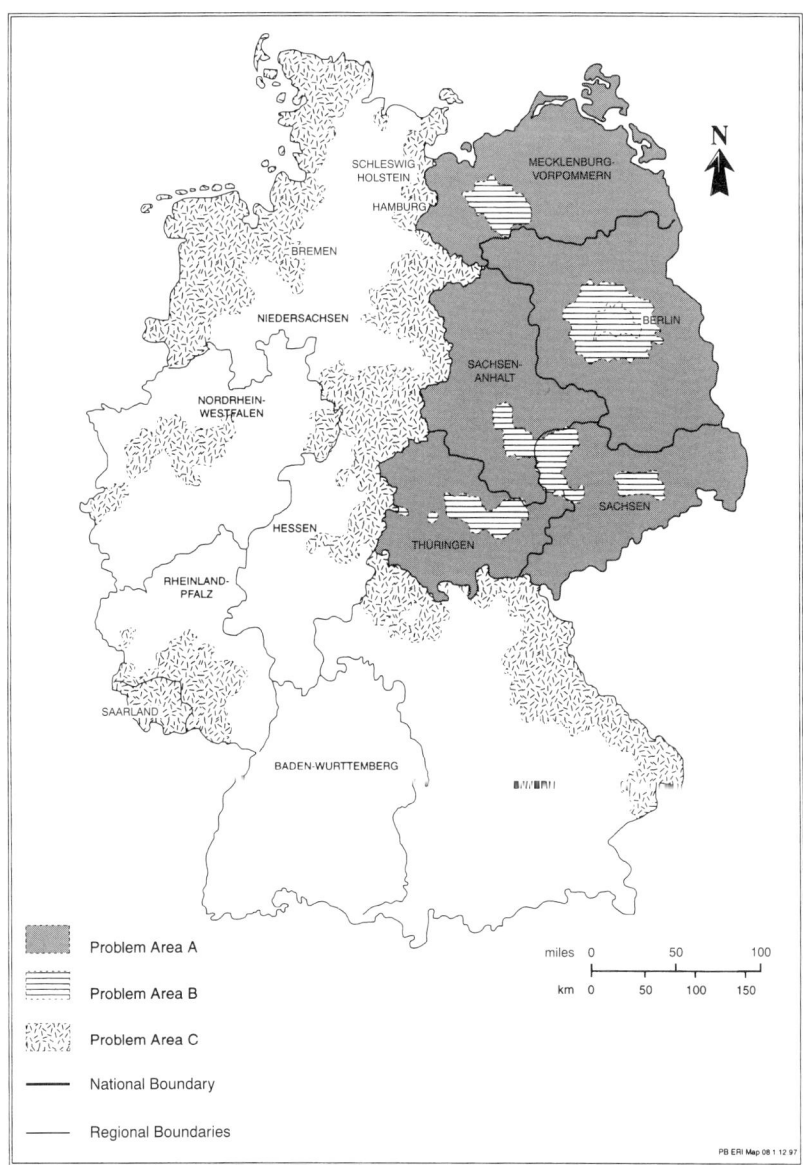

Figure 8.3 Government assisted regions, Germany, 1977–99

of the EU for the period 1994–99, it was not surprising that since economic development in the eastern *Länder* was lagging behind the west, the east received by far the largest share of Germany's allocation. Brandenberg, Mecklenburg-Vorpommern, Sachsen, Sachsen-Anhalt and

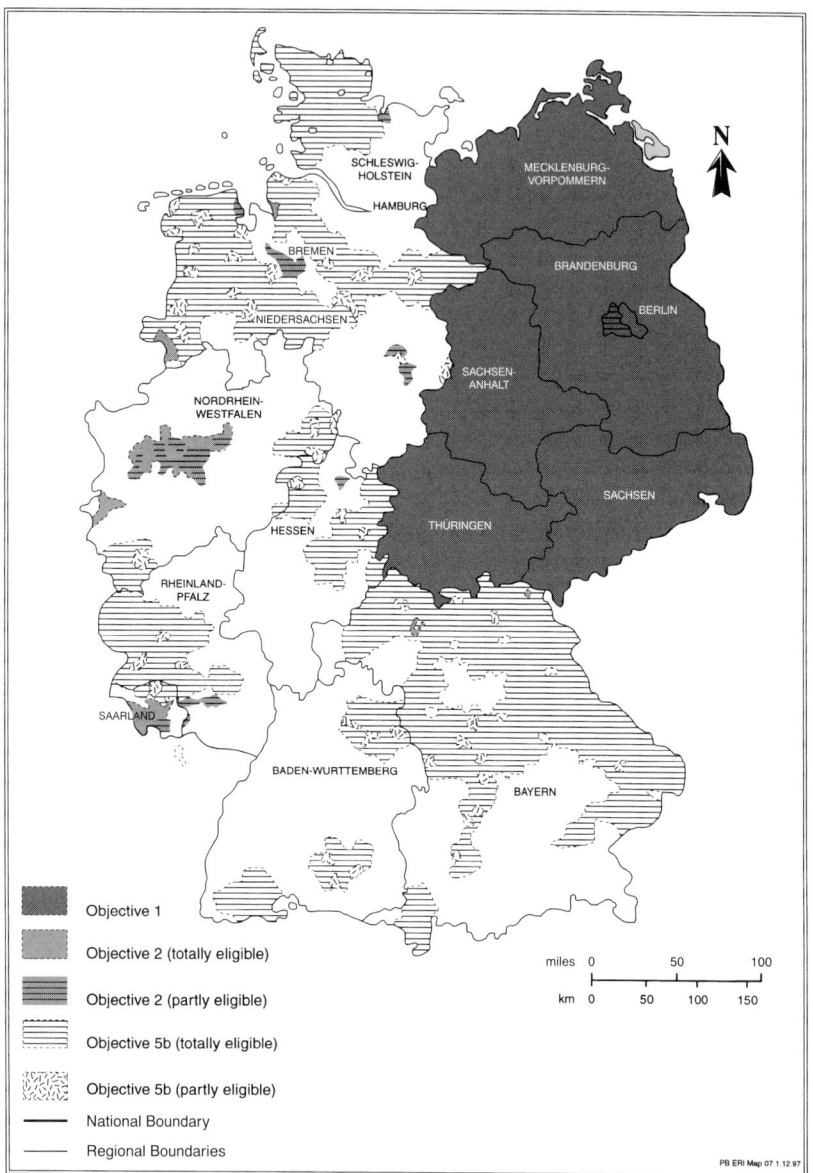

Figure 8.4 Designated Structural Fund areas, Germany, 1994–99

Thüringen, together with Berlin (all NUTS-2), were hence eligible for Objective I funds of up to 13.6 billion ECU, 1994–99, to help facilitate development and structural adjustment (see Figure 8.4). Taking into account the serious weaknesses of the East German economy, six priorities for funded expenditure were established in agreement with the Commission (Williams, 1996):

1   Measures to combat unemployment, and to facilitate human resource development and vocational training (27.3 per cent).
2   Measures to promote agriculture and the development of rural areas and fisheries (24.2 per cent).
3   Productive investment (18.2 per cent).
4   Financial and service support for small and medium size enterprises (17.4 per cent).
5   Improvement and the protection of the environment (8.3 per cent).
6   Research, innovation and technological development (4.6 per cent).

The comparatively small volume of Objective 2 funding, amounting to 1.6 billion ECU, 1994–99, was allocated primarily to the *Regierungsbezirke* of the Ruhr coalfield area (NUTS-3 areas) to help facilitate the economic conversion of an area still affected by industrial decline, and Objective 5b funds of 1.2 billion ECU were allocated to the rural areas of eastern and Alpine Bayern, the North Sea coastal lands and Rheinland-Pfalz.

Contrary to the situation in all other member countries of the EU except Belgium, eligible NUTS-2 in Germany (the *Länder),* are, because of their extensive powers of devolved government, authorised to submit their development plans directly to the Commission for consideration, rather than indirectly through central government. In contrast, eligible *Regierungsbezirke,* are obliged to submit their development or conversion plans to their *Länder* in the first instance (rather than to central government), prior to seeking funds from the Commission. Objective I areas, comprising the whole of East Germany, contained 20.7 per cent of the total Federal population, while only 8.8 per cent and 9.6 per cent of the Federal Republic's population lived respectively in Objective 2 and Objective 5b areas (CEC, 1996b).

Belgium

In the late twentieth century, the tightly organised central state structure of Belgium evolved into a Federal state as a result of the very great cultural, social and economic differences in the country and a succession of constitutional reforms (in 1970, 1980 and 1988). At the upper level, the central government retained responsibility for foreign policy, defence, a vestige of competence for monetary and fiscal policy and responsibility for foreign policy, defence and matters relating to the whole country such as the motorway network, but most other responsibilities, including planning, were almost entirely devolved to middle tier authorities.

The middle level of government, which has been granted a considerable degree of legal, political and administrative autonomy, is based on three cultural communities and three economic regions (Wiehler and Stumm, 1995). The Flemish, French and German-speaking communities have responsibilities relating to the use of language, the education system and cultural and personal affairs – including social policy, while the economic regions (comprising Vlaams Gewest, Région Wallonne and Bruxelles), have powers relating to employment policy, public investment, motorways and highways, regional policy and regional planning, and are thus responsible for formulating legislation on planning and producing regional and sub-regional plans.

The special law of January 1989 ensures that the cultural communities receive a substantial share of value added tax (accounting for about 80 per cent of their budget resources), while the economic regions derive about the same proportion of their budget resources from personal income tax – revenue being transferred from the upper to both middle tiers of government on a pro-rata basis. The regions can also raise their own tax revenue (for example, motor vehicle tax and surcharges on various taxes). Clearly, the tax revenue base of the economic regions helps regional policy and planning.

At the lower level there are nine provinces and 589 municipalities, but the responsibilities of the provinces have been downgraded by the recent reforms and their powers substantially subsumed by the economic regions (the provinces accounted for only 2.5 per cent of Belgium's total public spending in the mid-1990s), while the municipalities look after matters of local interest such as the local road network and local utilities. Most of their budgetary resources are derived from central government allocations and local surcharges on property, incomes and road taxes (Wiehler and Stumm, 1995).

Until the 1980s, the principles of the Belgian planning system were based on the Urban and Regional Planning Act 1962. A hierarchy of development plans subsequently emerged – a national plan (which was never implemented), regional plans, sub-regional plans (*Plans de Secteur*), and municipal plans (comprising *Plan général d'aménagement* and *Plan particulier d'aménagement*) (EC, 1994c). However, many regional and sub-regional plans were not prepared, while other sub-regional plans and many municipal plans, although being prepared, were ineffectively implemented because of rigid procedures, changing priorities after plans had been drawn up, inadequate means of an application, and the lack of financial resources (EC, 1994).

Recently, the planning system has undergone reform in response to the constitutional reforms of 1980 and 1988, and is still in the process of transition. Each economic region is adopting its own autonomous planning system and following a discrete course of action but, in general, within each emerging planning hierarchy lower-tier plans are to be subordinate to those produced by the higher tier authority (Suetens, 1986; Newman and Thornley, 1996).

In the Vlaams Gewest, spatial structure plans (providing the general policy framework) and spatial implementation plans are being produced under the Planning Act of 1996 within a three-tier system of regional, provincial and municipal government. There is an emphasis on sustainable development and on the need to concentrate growth in the urban areas particularly Antwerpen, Bruxelles and Gent.

In contrast, a two-tier system (involving only the region and the municipalities) is being developed in the Region Wallonne. As codified and prescribed in the *Code Wallon de l'aménagement du territoire, de l'urbanisme et du patrimoine*, each tier has the responsibility of preparing a structure plan (*Plan regional d'aménagement du territoire* at the regional level and *Schème de structure* at the municipal level) and a detailed zoning plan (*Plan de secteur* in the region and *Plans particuliers d'aménagement* in each municipality), with the latter being legally binding (EC, 1994).

The Bruxelles capital region has already introduced an ambitious two-tier system of planning. Under it, both the regional and municipal authorities have the responsibility of preparing development plans *(plan régional de developpement and plan communal de developpement) and* zoning plans *(plan régional d'affectation des sols and plan communal d'affectation des sols),* with the plans approved by the municipalities having to be approved by the region (EC, 1994c). Whereas development plans provide a general policy framework, zoning plans specify land uses.

It is not clear, however, whether the above hierarchies will effectively facilitate the application of spatial policy to such key issues as the preservation of cultural and social balance, the further development of diversified economic activity and transport infrastructure, and the enhancement of the natural and built environment in order 'to accommodate social and economic progress without reducing the quality of life' (EC, 1994: 146). Neither is it clear how cross-regional planning issues (which will inevitably arise in the future) will be co-ordinated (Newman and Thornley, 1996). It might be necessary to confer upon the national state a specific co-ordinating role, or to establish cross-regional authorities for this purpose.

Regional policy and regional incentives

As elsewhere in Europe, in Belgium there is a marked north-south divide in the spatial economy, but, unlike the United Kingdom and Germany, the prosperous areas of the country are mainly in the north rather than in the south. Whereas Vlaams Gewest benefits from its major cities having a coastal location or easy access to ports, and is thus being well suited to the development of modem steelworks, car assembly factories, oil refineries and food processing plants, in contrast, Région Wallonne suffers badly from the decline in its traditional coal and steel industries.

In the early 1950s, high costs of production in the Sambre-Meuse coalfield of Région Wallonne were already making it uncompetitive, but

ECSC subsidies helped to sustain the industry through most of the decade until the crisis of over-production in 1958. Despite an increase in government aid, the coal industry continued to decline throughout the 1960s (Williams, 1987). There was simultaneously a decline of steel production in the region because of the increasingly comparative advantage of northern Belgium as a location of large integrated steel mills.

In response to the 1958 crisis, the Belgian government attempted to regenerate the economy of Région Wallonne. The Regional Development Act of 1959 provided financial assistance to help renew existing plant and to establish new firms, and a further and more generous Act of the same name in 1966 offered a variety of grants, loans and tax exemptions for firms locating in areas of potential, including the Borinage coalfield of Région Wallonne (Williams, 1987). But decline has been so rapid since the 1960s that, despite government aid, there have been substantial problems of readjustment throughout the 140 km of the Sambre-Meuse valley (Minshull, 1996).

Partly because of macro-economic constraints following the oil-price hikes of 1973 and 1978, and partly in response to popular demands, the central government devolved the administration and funding of regional policy in 1980 to the regions. Although new development zones were established in 1982 (with reviews every three years), the area of the problem regions was systematically reduced so that, whereas they contained 39.5 per cent of the national population in 1982, by 1985 their proportion of the population had fallen to 33.1 per cent. In 1992, economic regeneration suffered another setback as aid was further cut in Région Wallonne when new directives abolished interest subsidies.

By 1994, in comparison with averages for Belgium as a whole and the twelve members of the EU, Région Wallonne was clearly lagging behind economically since it had both a disproportionately high level of unemployment and disproportionately low GDP per capita. The level of unemployment in the region was 13 per cent, compared with 9.7 per cent in Belgium as a whole and 11.3 per cent in the EU, and whereas Belgium nationally had a GDP per capita 13 per cent higher than the EU, the comparable GDP per capita for Région Wallonne was 9 per cent lower (Office of National Statistics, 1996).

With the allocation of Structural Funds for 1994–99, it was therefore not surprising that the Région Wallonne (NUTS-1 status) became eligible for Objective 1 funding in respect of one of its most disadvantaged provinces, Hainault on the Borinage coalfield (see Figure 8.5). A sum of 730 million ECU was available to facilitate development and structural adjustment. The eastern industrial province of Liége (NUTS-2) and neighbouring provinces to the north became eligible for Objective 2 funding amounting to 342 million ECU to facilitate the economic conversion of industrial areas in decline, while 77 million ECU of Objective 5b funding was available to facilitate the development and strict adjustment of rural areas in the

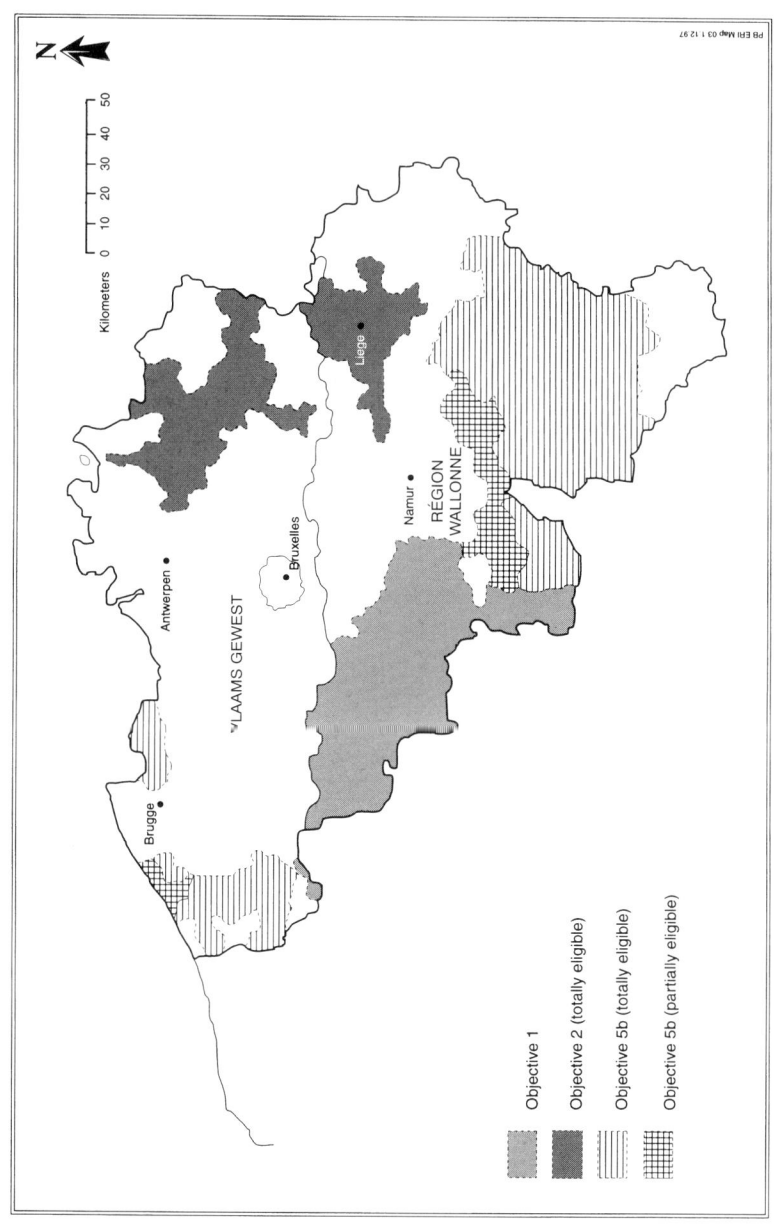

Kilometers
0 10 20 30 40 50

N

Antwerpen •

Brugge •

VLAAMS GEWEST

Liège •

Bruxelles •

Namur •
RÉGION
WALLONNE

Objective 1

Objective 2 (totally eligible)

Objective 5b (totally eligible)

Objective 5b (partially eligible)

Figure 8.5 Designated Structural Fund areas, Belgium, 1994–99

provinces of Luxembourg (in the Ardennes) and West Vlaanderen (in the north-west of Belgium). Because of Belgium's Federal constitution, Région Wallonne is authorised to submit development plans directly to the Commission in applying for funds, unlike regions elsewhere in the EC (with the exception of the *Länder* of the Federal Republic of Germany), although the provinces are required to submit their development and conversion plans to their regional administration for consideration before seeking funding from the Commission.

Areas eligible for Objective 1 funding contained 12.8 per cent of Belgium's population in 1994, Objective 2 areas contained 14 per cent, while only 4.5 per cent of the population lived in Objective 5b areas (CEC, 1996b).

9 Transition states of East Central Europe

Introduction: Communist legacy and transition

The contemporary local and regional planning and development policies in East Central European countries (the Czech Republic, Hungary, Poland and Slovakia) are influenced by the legacy of the Communist system, transition from centrally planned to market system, transformations in the local government system, disputes about the new role of spatial planning and the association agreements with the European Union.

Communist legacy

The Communist centrally planned system of allocation of resources has been characterised by a hierarchically organised system of national, regional and local planning. There was national and regional economic planning, and national concepts of settlement structure and physical planning on regional, urban and intra-urban levels. In regional economic planning, the spatial goals were governed by the national planning of the allocation of economic activities, labour force and housing. Regional plans were sums of spatial proposals of various ministries. The regional economic planning was supplemented by settlement development planning intended to govern the urbanisation process. No regional or intra-urban policies in the western sense were applied. The role of physical planning was to design a concrete spatial arrangement of objectives declared in economic development plans.

In the first decades, a national economic planning focused on massive industrialisation and sectoral economic decision-making was crucial for regional development. The allocation of investment to new industries usually reflected both the politically declared equalisation principle and economic principle favouring agglomeration economies. New industrial plants were established in backward rural areas creating single company towns, in newly established industrial towns and existing industrial centres. The industrialisation programme should have strengthened the economic base of Communist countries competing in the geopolitically polarised world and created modern urbanised society with equally accessible benefits to all citizens.

Since the 1960s, investment in the sphere of production was supplemented by consumption targets, namely in the sphere of housing development and provision of services in the system of selected central places. Standardised dwellings and services were provided across the country's territory. However, the top-down distribution of funds disadvantaged the lower ranked central places and non-centres. At an urban level, physical planning involved the use of elaborate and rigid land-use zone plans, which regulated the allocation of land for new housing and industrial construction.

The economic take-off of the post-war period ended in the mid-1970s. The earlier political rhetoric stressing the rapid growth has been transformed to preferences given to the qualitative aspects of development, increasing living standards of population and the consideration of environmental issues. Experimentation with the decentralisation of state power to local governments and private sector actors was applied in Hungary and Poland, while the Czechoslovak government kept strict central planning principles.

Transition

The post-1989 economic transformation turned earlier industrial strongholds into areas of comparative disadvantage, changed the relation between public and private actors in favour of the latter and cities and regions became areas for the location of private investment instead of objects of public planning (Lorenzen, 1996). The territorial development reflected burdens inherited from the Communist period as well as the new spatially selective activities of private investors. Inequalities increased with the decline in traditional industrial regions and foreign investment targeted on capital cities, selected regional centres and western border regions. New demands from market actors and newly created spatial inequalities become the basic contextual characteristics that should be integrated within a new spatial planning system.

The removal of Communist institutions was immediate, however, its replacement with a new system is a much slower and complicated process. The transition period is characterised by the political fight over the particular form of market-oriented economy. While the old principles were quickly rejected, new institutional arrangements are developing slowly and often in a chaotic manner creating many uncertainties. This applies in particular to fields such as planning, which has been regarded with suspicion and often seen as contradictory to the desired free market system.

The very liberal thinking of first transition years was characterised by low political priority of the central government given to physical planning, regional policy, housing policy, etc. (Sýkora, 1994b). Hierarchically organised economic planning was terminated, regional governments abolished or their powers strongly reduced and the responsibility for physical planning was devoted to newly established municipal self-government. The absence of comprehensive national spatial development strategy and consistent regional policy, changes in the local and regional government system and disputes about new planning legislation created contextual and institutional uncertainty.

Consequently, land-use planning on a municipal level and public regulation of development process were characterised by a preference for ad hoc political decisions to long-term strategic visions. In this situation ad hoc approaches have developed, with local governments applying their own strategies, often incorporating elements from before 1989 (Newman and Thornley, 1996). Importantly, physical planning on an urban level is being supplemented by an emergence of strategic planning and attempts to implement economic tools for stimulation and facilitation of local development.

The Czech Republic

The former Czechoslovakia was probably the country with the strongest equalisation policy in the Central and East European region. The Regional development in the years following the end of the Second World War was influenced by the attempts to repopulate the western frontier zone, from which about three million Germans were expelled in 1945–46, by the industrialisation of Slovakia through both the relocation of factories from the western frontier zone to the east and new inward investment to industry, and by flows of industrial investment to economically weak regions in the south of the Czech part of the country. Concerning the settlement structure, new industrial investment was targeted on small towns of 10,000 to 20,000 inhabitants (Musil and Ryšavý, 1983).

The main priorities of the 1950s and 1960s included the industrialisation of Slovakia and the concentration of new investment to heavy industries in existing industrial centres (Pavlínek, 1992). The equalisation policy on the macro-level and geopolitical reasons were the main ideological and strategic reasons behind the industrialisation of Slovakia, which was located in the middle of COMECON countries and remote from the western military borderland of the Warsaw Pact. Since the mid-1950s, the traditional industrial regions of North Bohemia and North Moravia were preferred to backward areas in investment decisions.

Until the 1960s, regional development was an outcome of a single national plan of economic development. Since the beginning of the 1960s, the national plan also included regional development projections and in the second half of the 1960s, the plans for selected urban agglomerations and frontier districts were added. The accent in the territorial distribution of resources was put on medium size cities as cores of regional agglomerations to utilise economies of scale. The reform process of the late 1960s brought an introduction of modern regionally selective industrial policy measures, such as investment grants and increased depreciation allowances, that were applied in selected regions (Blažek and Kára, 1992). The regional development priorities broadened through the inclusion of service and consumption targets. The long-term concept of national settlement network divided towns and cities into several hierarchical categories, each with a specific set of service functions to be provided for their regions. In the sphere of housing, the programme of construction included an important element to facilitate the regional distribution of the labour force. The equalisation process between the Czech lands and Slovakia came to a political phase by the establishment of a federation of two republics in 1968.

The normalisation process of the early 1970s strengthened the centrally organised system of the resource distribution. The border zone and selected industrial districts have enjoyed preferential treatment through the allocation of investments, investment grants, supplements to wages of selected professions (Blažek and Kára, 1992) and labour stabilisation housing allowances for private and co-operative housing construction (Sýkora, 1996). The concept of urbanisation and settlement structure has been advanced from the simple form of hierarchically organised nodes to the delimitation of regional agglomerations, urban regions and other central places. The concept was intended to manage and control the process of urbanisation until 2000. It influenced the distribution of resources with a strong preference given to agglomeration economies.

The 1980s were affected by a general economic decline. The one-sided rhetoric about economic growth was supplemented by the attention given to social and environmental aspects. Following the adoption of the Regional Planning Decree in 1977 (Blažek and Kára, 1992) regional planning was integrated into the jurisdiction of regional and district authorities as a sub-system of central planning aimed at spatially rational and equal distribution of resources. The politically declared goals of newly constituted regional planning included the rational distribution of resources and the effective use of forces of production on the whole territory of the state, and optimal use of natural, social and economic conditions of regions with the objective to improve living standards of the population. The first regional plans were prepared at the end of the 1980s for selected regions, however, due to the political change in 1989 and the abolition of regional governments in 1990, their implementation was hindered (Pavlínek, 1992).

Physical planning operated especially on the local urban level. After the Communist take-over, physical planning instruments were not used and the spatial allocation of investments to construction was governed by individual political decisions. Physical planning was reconstituted as a tool for urban development only in the 1960s. Physical plans designed the macro-spatial structure of urban areas, their general land-use pattern and especially focused on the allocation of land for housing and industrial construction and transport network arrangement (for more details see Carter, 1979;

Rubenstein and Unger, 1992; Sýkora, 1995). The amount of services on the city wide and neighbourhood levels were planned according to nationally set standards. The protection of arable land and the preference for high-density high-rise housing estates on city outskirts led to the creation of compact urban structures and limited urban sprawl. Physical planning was directed mainly at the implementation of construction targets set up in economic plans and the elaboration of spatial frameworks for future investment planning.

The introduction of the market system brought an increase in regional disparities (Blažek, 1996). The growing differences in economic development supplemented by political disputes between political representations of the Czech and Slovak parts of the federation brought a split of Czechoslovakia into two independent states (in 1993). Territorial disparities also emerged in the Czech Republic itself with growing unemployment in old industrial regions and backward areas, more active development of entrepreneurship in the western part of the country and spatially selective concentration of foreign investment interests in the capital city of Praha, selected other large cities and towns and the western border zone.

Regional development has not been influenced by any consistent approach of the central government. The intervention into spontaneous development has been considered as inconsistent with the market system. Ad hoc programmes were used for regional crisis management, and individual and unco-ordinated programmes with regional goals were introduced along with support to small- and medium-size enterprises, the labour market and agrarian interests. The emerging regional problems of the second half of the 1990s, with the rate of unemployment exceeding in some districts 10 per cent, and the pressure coming from the Association Treaty with the EU will probably result in the development of a more comprehensive institutional system of regional development policy.

The previous hierarchical system of national, regional and local planning was abandoned and there is no national or regional planning concept that would create a framework for the preparation of local physical plans. The very idea of planning has been treated with suspicion and one of the main tasks for planners has been to keep planning regulations in operation and defend the legitimacy of the planning system (Hoffman, 1994; Sýkora, 1995; Hammersley, 1997). Currently, physical planning and development control is characterised by the absence of national and regional spatial development concepts, unco-ordinated planning efforts of individual municipalities and by a strong pressure of various developers on weak and inexperienced local governments in attractive and valuable areas.

Territorial administration

The Czech Republic has a population of 10.3 million and a territory of 78,900 km^2. In 1990, the old hierarchically organised system of National

Committees, which represented state power in regions, districts and municipalities, was abolished and a new system of local government created by an amendment to the Constitution and through the new Municipal Act and District Office Act (Dostál *et al.*, 1992; Kára and Blažek, 1993). In Autumn 1990, for the occasion of municipal elections, Regional National Committees were abolished without replacement, District National Committees were replaced by District Offices that represent the state administration and municipalities became the basic units of local self-government. The present system of territorial administration consists of two tiers of seventy-seven districts and about 6,200 municipalities (Table 9.1). The capital city of Praha is a municipality which is further sub-divided into fifty-seven boroughs and thirteen so-called statutory towns are also divided into boroughs.

The new system of local government that has been in operation since 1990 is based on the separation of local self-government from state administration. The basic organs of municipal self-government are a directly elected Municipal Assembly and a Municipal Council and Mayor, elected by members of the assembly. The new Constitution, that was approved in 1991, also declared the existence of self-government with directly elected regional assemblies on the regional level. In 1997, after long political disputes, it was decided that fourteen regions will be established by 1. January 2000 (see Figure 9.1). However, the competencies of regions has not been specified yet. At the district level, there is a District Congress (Assembly), consisting of representatives of municipalities (often mayors) delegated according to the population size of municipalities. While large towns have many votes, there is a single representative for several small villages, a strongly biased urban-rural distribution of voting power. The District Assembly has very limited power and its role concerns the distribution of a central government equalisation grant from the district to the municipalities and the approval of the District Office budget.

The bulk of the state administration tasks is divided between seventy-seven districts and about 380 specially commissioned municipalities with delegated tasks of state administration. Municipalities itself are also responsible for certain state administrative tasks delegated to this level. However,

Table 9.1 Units of territorial administration in East Central Europe

Country	Regional and local administration		
	Regional	*District*	*Local*
Czech Republic	– (14, from 1.1.2000)	77	6196
Hungary	19 + 1	–	3130
Poland	49 (16, from 1.1.2000)	– (308, from 1.1.2000)	2459
Slovakia	8	79	2825

Source: Horváth (1997)

Figure 9.1 Approved regions (in operation from 1.1.2000) and existing districts of the Czech Republic

the range of delegated tasks differs, for instance building offices are localised only in some municipalities and provide their services for a set of surrounding municipalities. Municipal administration is in this field subordinated to the District Office and Municipal Assembly has no influence on the performance of delegated tasks (Perlín, 1996). It is envisaged, that in the long term, the level of fourteen regions and about 380 small regions should be strengthened at the expense of districts and small municipalities.

Local government in Prague is regulated by several legal documents (Blažek *et al.*, 1994, Kára, 1992). The Municipal Act declares Praha as a municipality and thus creates a background for the unified and centralised local government in the city. The Act on the Capital City of Praha from 1990 divides Praha into city parts (boroughs) with directly elected self-governments and their own budgets. The Charter (Statute) of the City of Praha is a local by-law, which specifies the deconcentration of responsibilities from the municipality (the central city government) to its boroughs. For instance, according to the Municipal Act, Praha as a municipality is the owner of real estate, however, it decentralises the management of certain properties, such as housing, to its boroughs. The Praha government administrate similar state functions as District Offices and delegate many of the tasks from the city level to selected borough governments, which serve population of their territory and adjacent boroughs (there are several levels of decentralisation). The major problems with local government in Praha are the large number of boroughs, the huge difference in their size and power (the smallest borough has less than 200 inhabitants while the largest has a population over 140,000, Blažek *et al.*, 1994) and the complicated system of decentralisation of state administration to the boroughs.

Regional policy

In 1991, regional policy was declared an integral part of general economic and social development policy with the main aim of providing the pre-conditions for the attainment of adequate working and living conditions of the population in all regions of the Czech Republic (Blažek and Kára, 1992). In the 1991–92 period, the government and Ministry of Economy in particular pursued a broad concept of regional policy. According to a set of criteria, there were recognised regions affected by structural change, backward frontier regions, regions with neglected infrastructure and regions and localities with environmental problems. In these regions, two-year tax holidays for private enterprises, grants for infrastructure improvement, support for active employment policy and some other measures were introduced (Kára, 1994). The new Regional Development Act, which was prepared in a draft version in Spring 1992, was refused by a new government after parliamentary elections in 1992 and the government cut former funds allocated for assistance to regional development.

The government resolution concerning regional economic policy (approved at the end of 1992), became the basic document for the realisation of regional policy and operational throughout the rest of the 1990s. The support is limited to small- and medium-size enterprises (SMEs) in areas delineated annually according to the unemployment rate. The areas account for 20–25 per cent of the Czech Republic population. The state through the Czech and Moravian Guarantee and Development Bank provides guarantees for loans and interest subsidies. The incentives should in particular support job creation and the export capabilities of firms. The programme of regional assistance is additional, i.e. the firm should first qualify for one of the basic programmes within the general support provided to SMEs and if located in an assisted region can apply for additional support.

In 1994, a specially designed programme was applied in four districts with the highest unemployment. The package of incentives contained support to SMEs, development of entrepreneurship in agriculture, municipal and transport infrastructure, ecological investments and active labour policy. The unemployment rate in the assisted districts fell sharply, however, it generally dropped in the whole country and thus the contribution of the regional help package cannot be evaluated precisely.

Regional policy (together with physical planning and housing policy) is the responsibility of the Ministry of Local Development. Up to now, there have been no new regional policy programmes designed and applied by this Ministry. Therefore, regional policy has been characterised by an ad hoc approach in the case of crisis management and limited support to SMEs. In 1997, the Ministry prepared the principles of regional policy, however, the regions are not at the top of the political agenda. Nevertheless, it might be expected that the duties emanating from the Association Agreement with the EU will change the government's perception of the role of regional policy.

There are other ministries and government agencies whose programmes include important regional policy elements. Probably the highest impact on overall regional development has been the system of local government finance (Blažek, 1994a, 1994b, 1996, 1997b; Surazska and Blažek, 1996) and the distribution of equalisation grant in particular. The Ministry of Labour and Social Affairs pursue an active employment policy which – through the network of labour offices – is targeted namely at districts with a high level of unemployment. The Ministry of Agriculture has developed the Programme for Rural Revitalisation that addresses the development of infrastructure in villages, revitalisation of rural built environment and public green spaces in villages. The Support and Guarantee Fund for Agriculture and Forestry allocates within the Agroregion programme, an additional support to farmers who have already received finance from one of the basic programmes of the Fund. The Ministry also supports reforestation in mountain areas and the preservation of cultural landscapes in rural areas. Transport infrastructure and environmental investment have also been to a limited extent influenced by principles of differentiated regional allocation with preference given to remote areas and regions of severe environmental damage. The Ministry of Culture supports conservation and regeneration activities in protected historic settlements.

The Ministry of Trade and Industry established two agencies that have a strong influence on local and regional development. CzechInvest is an agency for the support of foreign investment. This agency co-operate with various local actors, especially local governments in towns and cities and their departments of urban development and physical planning and with Regional Development Agencies. The agency is involved especially in consultancy and organisation of real estate provision to potential foreign investors. It has also organised a programme of accreditation for towns and cities which offered training in local economic development practices. The Business Development Agency was established by the Ministry with the assistance of the PHARE programme. The agency created a network of Regional Advisory and Information Centres, aimed at providing consultancy to SMEs, and Business Innovation Centres, akin to science and technology parks.

Two regional development programmes have been created for areas heavily affected by industrial restructuring. The preparation of regional development programmes for Ostrava and Northern Bohemia have been sponsored by the PHARE programme that also co-financed the establishment of Regional Development Agencies in these areas. Since 1994, the PHARE CBC programme supports cross boundary co-operation between the Czech Republic and Germany and since 1995 with Austria as well. The PHARE programme is important, not only because it provides investment grants, but also because of the know-how transfer which takes place through the application of procedures used in the European Union.

Policies of regions

The formation of independent policies on the regional level is severely restricted by the non-existence of self-government at the regional level. This should change with the introduction of regional government in the year 2000. Up to now, a very limited role has been played by District Offices, which beside their administration responsibilities also attempt to substitute for the non-existence of self-government at this level and have been engaged, for example, in the promotion of the district in the sphere of tourism, etc. Since the abolition of regional government at the end of 1990, there have been selected attempts to co-ordinate some activities at the regional level, of which the most important have been the establishment of Regional Development Agencies.

District Offices are directly subordinated to the Ministry of Interior and its departments to other ministries. Their role in the local development is limited to management of hospitals, social care facilities, libraries, museums, theatres, etc., which have not been transferred to municipalities. There are departments of regional development within District Offices. They often organise and finance the preparation of physical plans for municipalities, despite the fact that this task should be carried out at the lower level of specially commissioned municipalities with delegated tasks of state administration. There are many cases when these departments order and pay for the preparation of a district development plan, despite there being no legal requirement for such a document and no self-government body which could pursue its application. The role of overall development planning at a district level is not even performed by the District Assembly. Its influence on the redistribution of the central government equalisation grant to municipal budgets can have very limited implications for the development of the district, for instance in the case of reserving part of the grant for investment in a common technical or service infrastructure project.

Since 1993, several Regional Development Agencies have been estab-lished by various local institutions, including towns, local enterprises and banks, municipal associations, trade unions, etc. They are independent bodies whose role, is not regulated by the state. The central government has been involved only in the founding of the first North Moravian RDA in Ostrava to tackle problems of the old industrial and coal mining region (it includes six districts). The second North Bohemian RDA in Most was established in another old industrial and lignite mining region (seven districts). Since 1996, other RDAs emerged, for example, in the central

Moravian town of Olomouc. The RDAs were originally created as institutions for gaining grants, subsidies and other forms of financial help to the region and in particular for institutions that established them. At present, they act mostly as a consultancy service for both local governments and the private sector. Their revenues come from the support allocated by shareholders, consultancy services and grants from the PHARE programme. The most active is RDA in Ostrava, which benefits from the government and PHARE support. It developed a strategic plan for 1997–2000, that includes investments and subsidies to regional and local infrastructure projects and dissemination of regional information and propagation materials, and is involved in the EU ECOS/OVERTURE programme.

Local (municipal) development practices

Municipalities have a right to manage municipal property, adopt municipal budget, establish legal entities, adopt a municipal development programme, approve local physical plans and issue municipal ordinances. The basic local development planning documents declared in the Municipal Act of 1990 are the municipal development programme, that specifies long-term priorities of socio-economic development, the medium-term physical plan and the municipal budget, that specifies financial and in particular investment allocation in the short-time perspective. While budgets are necessary for municipal governance and physical plans are commonly used instruments, municipal development programmes are rarely adopted. There is only a small number of cities and larger towns, which are currently preparing municipal development programmes, often called strategic plans. Unfortunately, the Municipal Act is the only legal norm where municipal development programmes are mentioned and there exists no rules or guidelines for their preparation. Municipalities have to take their own initiative and experiment with the preparation of such planning documents. Up to now, the short-term individual and ad hoc political decision-making was preferred to long-term comprehensive strategies of local socio-economic development.

The Municipal Act of 1990 allowed for the disintegration of municipalities amalgamated during Communism. Consequently, the number of municipalities increased from about 4,100 in 1990 to about 6,200 at present (Table 9.2). This process led to an emergence of a large number of very small municipalities (about 60 per cent of municipalities have less than 500 inhabitants and a further 20 per cent of the population between 500 and 1,000). The self-government of such small municipalities is very weak in financial and professional matters and has limited bargaining power in relation to the state government as well as private sector developers. In many cases, small municipalities create associations and establish companies to organise certain tasks, such as the collection and liquidation of municipal waste or water, sewage and other technical networks construction and management.

Table 9.2 Number of municipalities and their average population

	Number	*Average population*
Czech Republic	6196	1667
Hungary	3130	3315
Poland	2459	15623
Slovakia	2853	1845

Source: Horváth (1997)

The main trend in municipal finance has been the decrease in the dependence on central government grants and the increasing role of revenue from an apportionment of personal income tax from individual entre- preneurs and employees, together with property tax and other own incomes, including local fees and revenues from the sale and lease of municipal property. Municipalities are also entitled to borrow money and issue com- munal bonds (this approach has been used, for example, by the capital city of Prague to gain finance for investment in transport infrastructure). There are large differences of own incomes per capita between municipalities (this is partly diminished by the central government equalisation grant). Surazska and Blažek (1996) indicate a regional pattern of this inequality with highest incomes achieved in cities and in the western part of the country, especially along the boundary with Germany and Austria. The system of local government finance has changed several times during the 1990s (Blažek, 1994) and this resulted in instability and caused difficulties for financial and investment planning at the municipal level. An important characteristic of municipal finance from the point of view of local develop- ment is that investment accounted for a high share (35–40 per cent) of municipal expenditure.

Physical planning and the control of development process

The regulations governing territorial planning and the control of the development process in the Czech Republic are provided in the Act on Physical Planning and the Building Act of 1976. New laws which reflect changing conditions are (at the time of writing) under discussion in the Parliament committees. Physical planning is in the competence of the Ministry of Local Development. The principal instruments of physical planning include planning working papers, planning documents and plan- ning permits. The purpose of the planning working papers is to collect basic data and evaluate proposed developments. Planning documents are the real physical plans, which differ according to time horizons (projection, plan, action project) and spatial scales (regional, urban, urban zone). In the proposed spatial planning act, the time horizons are abolished and planning documents can have the form of a regional plan, a general land-use plan for a municipal area and local regulation plan for a settlement zone. The planning permit is an executive decision of the state administration about the location of new development, land-use changes, the declaration of a protected area or the construction closure of a particular area.

The principal authority responsible for procurement of physical planning documentation is at the municipal level. However, for many small municipalities the preparation of physical plans is organised by District Offices or commissioned municipalities with delegated tasks of the state administration. The physical plans are approved by Municipal Assemblies and are binding at the lower levels of planning, and in respect of the elaboration of development projects and decision-making concerning the issue of planning permits.

The proposed spatial planning legislation concerns the organisation of the planning institutional framework on three basic levels. The central government will prepare the programme of national development. Regional governments (in operation from 2000) will prepare regional development programmes and regional physical plans, which will specify especially the organisation of regional transport and technical infrastructure and delimit the protected environmental zones. The regional governments will also co-ordinate the harmonisation of municipal physical plans. Municipalities will be the core institution of physical planning. The principle planning documents will be the municipal development programme, the land-use plan for the whole municipal territory and the detail regulation plan for an urban zone. In the case of small municipalities, land-use and building regulation principles will be applied in a single plan.

At present, general land-use plans are the most common planning documents and many local governments, especially urban and suburban, have organised the preparation of new land-use plans recently. The preparation of physical plans of neighbouring municipalities is unco-ordinated because of the absence of regional physical planning. The preparation of plans for small municipalities in suburban and other attractive areas is often strongly influenced by the pressure from developers. The preparation of new regulation plans for urban zones have been rather neglected. They are missing especially for areas with high development pressure, such as the central city of Praha (Sýkora and Šimoníčková, 1994). Unfortunately, local politicians preferred ad hoc decisions to long-term strategic visions of the urban development. Regional plans have been elaborated only in the 1980s (an exception was the Regional Plan for the Praha-Central Bohemian Agglomeration approved in the mid-1970s) and they have not covered the whole territory of the country. At present, there are no regional authorities which could be responsible for regional planning.

In Praha, the draft of a new Master Plan is currently (1998) under negotiation. The old Master Plan from 1986 has been replaced by a provisional plan from 1994. The City Master Plan of 1994 is based on the 1986 plan, from which it takes areas with relatively fixed urban structures where major functional changes are not expected and declares them as stabilised zones (Sýkora, 1995). The stabilised zones cover about two-thirds of Praha's territory and serve as a binding document for the preparation of local regulation plans and for the planning application procedure. The developments proposed in non-stabilised zones require preparation of detailed planning documentation (urbanistic studies), financed by the developer. The new Master Plan and the plan of stabilised zones use a principle of mixed zoning, that has replaced the monofunctional zoning used by physical planners in previous decades (Sýkora, 1995).

The development process is regulated in two steps: through planning application procedure and building application procedure. The responsible authorities are building offices (over 400 in the country). The authority checks if the application is in accordance with the approved planning documentation and the requirements of various state administration departments and organisations which are in charge of technical and transport infrastructure. It also organises public hearings to reach a compromise between different opinions on the development proposal. In protected historical urban areas, new developments are carefully checked by the historical monuments protection authority. Environment Impact Assessments is organised for industrial, trade and storage complexes with development areas in excess of 3,000 m^2. If the application corresponds to the requirements of the Building Act, the planning permit should be issued within 60 days from the date of submission. The permit is valid for a two-year period.

In the building application procedure a detailed plan of the constructed building is checked by building offices. The building permit can be granted only to those who have already obtained the planning permit and can prove the ownership rights. The application must contain approvals and statements from several institutions, such as the hygienist office, utility companies and departments of local administration. The processing period should not exceed two months. Building permits entitle the recipient to commence the construction work. They lose their validity if the construction work does not commence within two years from the date of issuing the permit. After the completion of a building, a certificate of approval must be issued by a building department for the building use and occupation.

Hungary

Hungary was, in contrast to other communist countries, characterised by gradual reform, decentralisation of decision-making, experimentation with new models and the small, but important role of private and shadow economy. After the communist take-over, centralised national planning was constituted as a crucial means of economic management of the country. It was based on hierarchically organised top-down relations in industry as well as local government. In 1968, the New Economic Mechanism, to an extent, decentralised decision-making and introduced a greater degree of flexibility at lower levels of the economic planning system, and the step-by-step reforms of the 1980s brought some elements of market system into the Hungarian economy (Lorenzen, 1996). Furthermore, the new Law on Councils of 1971 granted more autonomy to local government (Enyedi, 1990a).

After the Second World War, regional development was an outcome of national economic planning aimed at promoting industrialisation. Communists intended to transform rural agrarian society to an urban and industrial one by the means of industrialisation and collectivisation in agriculture (Zoványi, 1986). The policies also included the reduction of the dominance of Budapest. Socialist industrialisation emphasised investment in heavy industry. New large state enterprises were established in a group of new towns and some other existing settlements. However, in the 1950s industrialisation did not eliminate disparities between urban and rural areas.

In the 1960s, large investment projects focused on the five growth poles of Miskolc, Debrecen, Szeged, Pécs and Győr, that were designed as counter-poles to Budapest (Lorenzen, 1996; Zoványi, 1986, 1989). The industrial dispersion policy also promoted the development of light industries in small urban centres and backward areas. Furthermore, the purposeful relocation of enterprises from Budapest, the preferential treatment for the location of enterprises outside of Budapest and the development of industrial enterprises associated with agriculture also contributed to a more balanced regional pattern.

In 1971, the government adopted the Concept of National Settlement Network Development (Horváth, 1995; Zoványi ,1986, 1989). The Concept which outlined the development of settlement structure up to the year 2000 was based on the hierarchical model of central places. Nine hierarchical categories of central places were identified, including the capital, regional centres, sub-regional centres and local centres. The rank of centres was defined by the functions and services provided by the centre for its region. The ranking of settlements influenced financial flows to infrastructure, housing and services. However, the downturn in the Hungarian economy during the late 1970s restricted the original goal of even development of services provision across the country. Larger settlements received most of the finance and, consequently, the changes in the settlement system were characterised by the growth of larger towns on the one hand and the depopulation of villages on the other (Tóth, 1993). The discussion about the prevention of unnecessary out-migration from rural areas (Zoványi, 1986) influenced changes in the settlement policy. In 1985, a new programme of The Long-term Tasks of Regional and Settlement Development was approved with priority given to the co-operation between settlements, to the development of backward rural areas and to the protection of the environment.

Transformation and the introduction of the market economy brought an increase in regional disparities. On the one hand, there have been areas with concentration of foreign investment (80 per cent in the western part of the country and the vicinity of Budapest; MERP, 1996), and on the other hand, the decline has influenced rural areas and regions affected by de-industrialisation. The polarisation between Budapest and the rest of the country and the decline of wealth from west to north-east characterise the spatial pattern of uneven spatial development in the 1990s.

In the first period of political and economic transition, the regional development planning was not considered as a relevant policy instrument and regional development was without any regulation (MERP, 1996). The development of a new societal system to higher complexity and maturity and emerging regional problems have been basic contextual characteristics behind the development of the new institutional system of regional planning and regional policy that came into operation in the second half of the 1990s.

Territorial administration

Hungary has a population of 10.3 million and a territory of 93,000 km^2. The Hungarian Republic is divided into the capital, nineteen counties, twenty towns of county rank, 148 towns and 2905 villages (Hajdú, 1993). The capital is further sub-divided into twenty-three districts and towns may also choose to be divided into districts. The old hierarchically organised model of councils (local organs of state power and administration) was abolished in 1990 by modifications in the Constitution which were further elaborated in the Act on Local Self-Government.

There are two basic levels of local self-government: municipalities (towns and villages) and counties. The capital city with districts is a specific case, which will be described later. The responsibilities of local government vary, although each authority enjoys equal basic rights and there is no hierarchy to subordinate any one to another (Hajdú, 1993). However, while the Act on Local Self-Government brought independence and autonomy to municipalities it strongly reduced the functions of counties. The counties have only a subsidiary status and a county can assume only those functions which municipal self-governments cannot perform or refuse to assume (Pálné Kovács, 1993).

The mean population of a county is 524,000 (Surazska *et al.*, 1997) and the average size of municipalities is 3,315 (Horváth, 1997) (Table 9.2). Villages are smaller settlements with populations below 10,000. Towns are divided into two categories: towns and towns of county rank, the latter with populations over 50,000. The local government in a town of county rank, which is a municipal authority also performs functions delegated to the county. Consequently, these towns do not send representatives to the County General Assembly in the county where they are located and are therefore not part of the county's governmental responsibility.

County self-government is controlled by a directly elected County General Assembly (until 1994, the representatives were delegated by local government). The state interests at the county level are represented by prefects appointed by the President on the recommendation of the Prime Minister. The most important task of the prefect is the legal supervision of local government. In recent years, there has been discussion about the establishment of six larger regions that would comply to European Union territorial structures (see Figure 9.2)

The administration of the capital and its districts is regulated in a separate law. Budapest (population 2 million) has a two-tier administration. There are twenty-three districts with directly elected representations which form the basis of the city's self-government. The eighty-nine member city council of Budapest consists of both representatives of district councils (twenty-three seats) and directly elected representatives (sixty-six seats). The law provides both levels with equal legal status, there is no hierarchy and subordination of one to another. This offers the possibility of free bargaining between the districts and the capital (Hajdú, 1993). The common interests of the capital are usually of secondary importance in comparison with district matters. Consequently, the co-operation between autonomous districts and the capital has been increasingly difficult (Douglas, 1997).

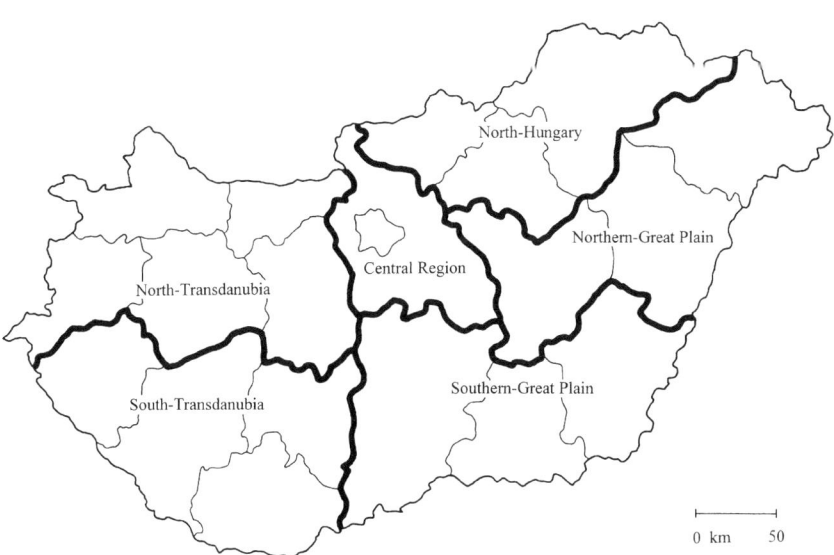

Figure 9.2 Proposed regions and existing counties of Hungary

National regional policy and regional planning

The primary role in regional development programmes in Hungary has been played by the Ministry of Environment and Regional Policy which was established in 1990. The main financial tool of regional policy, the Regional Development Fund (RDF) was founded at the beginning of the 1990s. In 1993, a decree on the principles of subsidies for regional development specifying the main tasks and means of regional policy was approved by the Parliament. The goals included regional crisis management, the economic restructuring of depressed and backward regions, the implementation of selective infrastructure projects, especially in backward areas, and national and international co-operation (Horváth, 1995; Lorenzen, 1996). The main task of the Regional Development Fund was defined as investment aimed at job creation, infrastructure investment and support to businesses in underdeveloped regions and regions with a high level of unemployment.

The areas that received assistance from the RDF (0.3 per cent of GDP in 1991–95) were 1325 small rural settlements that accounted for 17.4 per cent of the country's population (Horváth 1995). Towns affected by de-industrialisation have not been the subject of assistance. Most of the RDF finance was allocated to counties in the north-east of Hungary, more than 70 per cent of funding was spent on infrastructure development projects, such as gas, telephone, road, drinking water and sewage systems, and support for job creation projects was of minor importance (Horváth, 1995; MERP, 1996).

Further financial assistance came from the EU PHARE programme. It was used for the development of regional policy according to EU standards, providing assistance to two depressed regions suffering from the crisis of metallurgy and agriculture, and assistance to municipal associations to promote co-operation between small local authorities. The project aimed at problem regions ended in 1996 and, currently, the PHARE CBC (Cross Boundary Co-operation) programme assists an area bordering the Austrian backward region of Burgenland, thus channelling finance to one of the most developed regions in Hungary (source: correspondence with Gyorgyi Barta).

Horváth (1995) assesses that during the first half of the 1990s any clearly specified concept or strategy of regional policy was not formulated and initiatives were ad hoc, reactive and unco-ordinated. While the RDF was used in backward regions, the sources for modernisation of the public road and railway networks, investment concerning environmental protection, assistance for the industrial development, subsidies to agriculture, etc., were channelled to the most developed areas of the country (MERP, 1996). A change in the overall concept and harmonisation of development effort between the various ministries and other state agencies has been brought about by a new Act on Regional Development and Physical Planning which was approved in 1996. The law determines the rules and tasks of regional development policies and physical planning at national and regional levels. It is designed according to the principles of European Regional and Spatial Planning Charter and the EU regional policy. The main objectives of regional development policies and physical planning are to encourage development in every region of the country, to reduce differences between the capital, towns and villages as well as between developed and backward regions and to help the harmonious development of spatial and settlement structures. Tasks of regional development include assistance to backward regions and regions affected by economic restructuring as well as assistance to regions of high priority (development poles), the improvement of the conditions for innovation in settlement centres and the creation of a favourable environment for investors. Physical planning on national and regional levels determines the structure of land-use and rules of land utilisation, the spatial structure and location of infrastructure networks while taking into account the protection of the natural environment.

The law states that regional development and planning should be carried out in co-operation with the state, local government, economic and other interested organisations and individual persons. The basic documents at the nation-wide level include the National Regional Development Concept (which is set out in six-year periods and approved by Parliament), national physical plan and plans for a region of high priority (certain elements of these plans are statutory and are binding for local self-government and local physical planning), the principles of regional development support, and the criteria for the classification of eligible regions.

The preparation of the National Regional Development Concept started in 1995 and has been based on sectoral conceptions prepared by individual ministries. The Concept itself includes long-term objectives that concern major demographic, settlement and environment changes and the development of major infrastructure networks, and medium-term objectives that specify the regional allocation of the development programmes of individual ministries. There are three dimensions of spatial development policy on the national level (MERP, 1996): first is the determination of problem regions according to operational principles of EU Structural Funds, second is the preference for elaboration of regional strategies and programmes on the level of six large regions and third is a differentiated approach according to settlement size and function. The key principles of financing regional development specified in the National Regional Development Concept (MERP, 1996) include the decentralisation of resource distribution, the concentration of resources on the most important issues, the mobilisation of outside resources, the balance of the normativity and discretionality, and the promotion of complex development in the larger regions.

The law determined the establishment of the National Council for Regional Development that consists of representatives from central government, the Budapest government, County Development Councils, national chamber of commerce, employers' and employees' organisations and a

national representative of municipal associations. It should act as the recon-
ciliation centre for the different views of regional development held by the
involved institutions (MERP, 1996). The main role of the council is to
assist the government by making comments, proposals and evaluations of
proposals, principles, concepts and the implementation of regional development and planning programmes and policies.

Policies of regions

At the regional level, there are a number of actors involved in regional
development and planning. The counties have a duty to maintain those
services and institutions whose impact extends over a larger territory and
cannot be managed from the municipal level, such as county archives,
museums, theatres, social institutes for children, etc. (Lengyel, 1993).
They can also undertake any type of public duty which is not in conflict
with the interests of municipalities (Hajdú, 1993). However, counties have
very limited power, influence and financial resources and they play a role of
subsidiary administrative units (Horváth, 1995).

The tasks of county self-government declared in the Act on Regional
Development and Physical Planning of 1996 include the preparation of
physical plans for the whole county and/or its sub-regions and the co-
operation with the self-government of cities with county status in order
to co-ordinate physical plans for the city and surrounding area. Regional
development tasks within the county are co-ordinated by the County
Development Council (CDC) which is established and funded by the central
government, a respective county self-government, a county chamber of
commerce, a county labour council and local municipal associations. This
should promote networking among different county institutions. The president of the CDC is at the same time the president of the County General
Assembly. The CDC elaborates and approves the long-term regional
development concept of the county, the regional development programme
of the county and individual sub-programmes. The county physical plans
and objectives of the county regional development concept, which shall be
binding for municipal self-governments are approved by the County
General Assembly.

According to the Regional Development and Physical Planning Law of
1996 County Development Councils may set up Regional Development
Councils (RDCs), institutions whose task is to integrate development across
several counties (the central government strongly argues for the creation of
six regions that would comply with EU priorities). The Law defined two
mandatory councils: the RDC of Budapest and its Agglomeration and the
Balaton Development Council in the recreational area around Lake Balaton.
The RDCs consist of representatives from CDCs, the central government
and certain interest groups. They should participate in the preparation of
the National Regional Development Concept, help to integrate county

regional development concepts, bid for the state regional development resources and distribute them, etc. However, the relationship between the Regional Development Councils and local (municipal and county) self-government is not clarified in the law. For physical planning tasks at this level an institution of the regional Chief Architect is established by the central government to supervise the preparation and implementation of physical plans at both county and municipal (settlement) levels, initiate modifications and to comment on the National Regional Development Concept.

The basic documents on a regional level are regional development concepts, regional development programmes and physical plans. The regional development concept determines the long-term development priorities (over seven to fifteen years) and includes detailed concrete frameworks for medium and short-term planning. The development programme is a medium-term action plan and consists of strategic and operative programmes. The regional physical plan (or regional arrangement plan) determines the land-use pattern, spatial arrangement of technical and infrastructure systems and environmental protection.

Between 1991 and 1994, Local Enterprise Agencies were established in counties with the support of the PHARE programme and under the co-ordination of the Hungarian Foundation for Enterprise Promotion (Lorenzen, 1996). Their tasks are narrowly focused on small- and medium-size enterprises. The Chambers of Commerce with a compulsory membership were created by a law in 1994. They are organised on a territorial basis with chambers in each county. Their representatives are members of County Development Councils. Furthermore, the government may establish enterprise zones in regions undergoing industrial restructuring and municipalities and municipal associations may establish industrial parks and other development units to implement their specific regional development objectives.

Before the Regional Development and Physical Planning Law of 1996 came into operation, there have been individual cases in which regional development strategies were elaborated. Faragó (1994), for instance, informs about the South Transdanubian region where a regional development strategy was elaborated and the South Transdanubian Development Fund was established to serve regional development. The programme was launched due to financial support from several ministries. Founders included county and municipal governments and banks.

Local (municipal) development practices

The Act on Local Self-Government of 1990 granted municipalities (towns and villages) relative autonomy and financial independence. Municipalities consequently have the right to regulate and manage matters of local government, to own real estate and exercise property rights (there are some basic assets which cannot be sold, such as public spaces, utilities or certain buildings), to decide freely about their revenues, to have their own incomes and levy additional local taxes. They can establish businesses or participate in enterprises, and can approve rules that are not in conflict with higher-level regulations. The Act also defined the public duties of municipal governments that include: the development of the area; protection of the built and natural environment; housing policy; maintenance of the local road system and public spaces; public transport; water supply and sewage systems management; management of cemeteries; public order and safety; provision of kindergartens; primary education; social welfare and medical services, etc.

The Act abolished all the amalgamations among municipalities realised during Communism and now there are as many local governments as in 1949 (Enyedi,1994). Consequently, 35 per cent of more than 3,000 municipalities have less than 500 inhabitants. These small municipalities have many duties but little revenue (Lorenzen, 1996). Municipalities have the right to establish associations of representatives to tackle problems that cannot be solved by individual small municipalities. Usually, co-operation is achieved for matters concerning legal power, such as granting building permission, and the joint maintenance of institutions, such as schools and social care homes, are achieved. According to the Act on Regional Development and Physical Planning local governments can establish Regional Development Associations of municipalities and in co-operation with other legal entities.

The sovereignty of municipalities is restricted mostly by the system of local government finance. Despite local self-governments having the ability to levy local taxes, they usually do not use such instruments (with the exception of local business tax) and remain heavily dependent on the central government for their revenue (Alm and Buckley, 1994). In 1995, normative state support accounted for nearly 60 per cent of local budgets. Loan financing and sales of real estate are among the devices used for balancing local budget deficits. Municipal government is also increasingly interested in the possibility of using local economic development strategies to attract new businesses.

Physical planning and the control of development process

The new Act on Regional Development and Physical Planning from 1996 defined several spatial levels of physical planning: nation-wide, large regions (associations of counties supervised by Regional Development Councils), counties and small regions (voluntary association of municipalities). Regional physical plans are not legally binding documents. The new system of regional physical planning is now in its very beginning and it is difficult to evaluate its strengths and weaknesses. Physical planning at the municipal level, however, is regulated by separate legislation. Ordinary physical plans are prepared and approved by municipalities and are binding for the regulation of the development process.

The main regulations concerning the development process are specified in the new Building Act of 1996. Planning and building permits must be obtained for virtually all development. Permits are issued by specialised building departments of municipal authorities. The application must conform to the local land-use plan and the procedure involves a number of individual permits from organisations such as water, electricity and gas supply authorities, etc. Environmental impact assessment is required for large development projects, as defined in the Building Act. The protection and conservation of historical buildings is strictly regulated by the preservation authorities that are independent of local governments.

The 1980 Master Plan of Budapest concentrated on continued development of housing estates. It also reinforced the decentralisation of the central city to district centres. The 1988 Master Plan put an emphasis on rehabilitation and the growth of the inner city. In 1986, the Master Plan was supplemented by the plan of the metropolitan region. A new concept of urban development and a concept of a new Master Plan was in preparation in 1998. The concept of regional development for the surrounding county, Pest, was approved in 1997. The work on the new master plan is supported from the EU programme ECOS-OVERTURE.

Budapest districts have a large autonomy in decision-making, not least in the field of planning and development. The right to implement development priorities and zoning regulations is vested with the individual districts. The chief architect office of each district implement plans and policies that deal only with local matters while the relationship between the districts and the city as a whole remains unresolved (Douglas, 1997).

Poland

The post-Second World War modernisation of Polish society was grounded within the framework of socialist industrialisation, but beside the restoration of industrial production were attempts to level out regional differences. The industrialisation programme of the 1950s was based on the establishment and development of large enterprises to secure the economies of scale which was in contradiction with the declared goal of a more equal spatial distribution of production capacity. Consequently, a further concentration of economic potential in already developed centres and newly established towns in the industrial regions reinforced the existing pattern of urban settlement (Wecławowicz, 1996; Gorzelak, 1996). Nevertheless, some medium-size industrial plants were located in less developed regions (Regulska, 1987) and the degree of concentration of industrial production in the traditional core region in the south of Poland between Kraków, Łódź and Wrocław diminished from 60 per cent in 1950 to 36.4 per cent in 1970 (Wecławowicz 1996).

In the early 1960s, regional planning was formulated and regional planning offices established on a regional (voivodship) level. Regional planning was economic in nature, focusing on distributing investment to production, infrastructure, housing and services. Spatial planning elements were subordinated to economic goals. The development of the settlement structure was influenced by the priority given to medium-size towns, where lower development costs were expected, and to the deglomeration policies that for instance included relocation of plants from Warsaw to the surrounding region. Within cities, planners implemented the separation of industrial districts from residential areas and introduced the concept of neighbourhood units for the development of housing estates.

Socialist industrialisation was accompanied by a housing shortage and environmental pollution in urban areas. The National Plan of Spatial Development, which was adopted in 1974, aimed at raising living standards and satisfying the consumption needs of the population, and at the protection and more effective use of the natural resources (Regulska, 1987; Enyedi, 1990a; Wecławowicz, 1996). The Plan also defined a system of urban agglomerations as the basic element of the settlement network (Regulska, 1987). The spatial policy became a compromise between economic objectives working in favour of concentration and the political objective of more equal development. At the beginning of the 1980s, new acts on socio-economic planning and spatial planning were approved. Spatial planning became equal to economic planning, the hierarchical subordination of local to regional and national plans was replaced by a bargaining process between those levels, more attention was given to the participation of the population in the planning process, and the right to approve local physical plans was transferred to municipalities (Regulska, 1987).

A new period in the development of Poland came with transformation in the 1990s. The basic ideological assumption of transformation policies was that market mechanisms will replace the central planning system in the allocation of resources and that market forces should be the sole means of regulating of the economic system, including its territorial structures. Transformation policies were in their nature macro-economic and during the first years of transformation there was no place for regional policy. Actually, the neglect of regional policy can be treated as a specific type of policy itself.

The first years of transformation were characterised by widening regional disparities (Wecławowicz, 1996; Gorzelak, 1996; Paul, 1995). Market competition revealed the economic strengths of certain regions and exposed the weakest regions. The traditional industrial agglomerations of Upper Silesia, Wałbrzych, Łódź and a number of single company towns suffered from an economic crisis but new economic activities developed in other areas, such as Warszawa, Poznań, Gdańsk, Szczecin, Wrocław, Kraków and Bielsko-Biała (Kortus, 1996).

The spatial concentration of social and economic problems – and the end of the illusion that the invisible hand of the market will solve all problems – brought the first attempts to formulate and implement regional policy initiatives. Concerning the interest of central government in regional development there has been a change from the comprehensive and hierarchically organised distribution of resources based on long-term visions, to reactive, ad hoc and spatially selective central government policies focused on problem areas.

Territorial administration

Poland has a population of 38.6 million and a territory of 312,700 km^2. It is the largest country in East Central Europe with a population 50 per cent greater than that of the Czech Republic, Hungary and Slovakia in total. There are three tiers of government: central, regional and local. Since 1975, the territorial administration has consisted of 49 regions (voivodships, *wojewodztwo*) and about 2450 municipalities (*gminas*) (Table 9.1). On average, voivodships have a population of 800,000 and *gminas* 16,000 (Strong *et al.*, 1996).

Until 1990, the country was centrally administered. The Local Self-Government Act of 1990 granted complete autonomy and delegated certain rights and responsibilities to municipalities, such as the right to own property, collect taxes, manage their financial resources and formulate and promote general municipal interests (Grochowski, 1997; Regulska, 1997). Municipalities are legal entities with directly elected councils and represent the interests of local community rather than central state administration. Regions are representations of the state and are subordinated to the central government. Each voivodship is administered by a governor (voivod) appointed by the prime minister. Parallel to the voivodship structure there exist voivodship assemblies, that consist of representatives from municipalities. Their power is limited, however, and they play an advisory role. They can raise issues with the voivod, supervise municipalities and mediate in conflicts between them (Regulska, 1997).

The territorial organisation of the state is the responsibility of the Council of Ministers Office which is in charge of the reform of the territorial administration and relations between municipal self-government and state authorities. The current Polish government is pursuing administrative reform, which would create twelve new regions. There are also other proposals, with the number of regions increasing to twenty-five. These regions would have substantial powers and responsibilities and could act as representatives of regional planning and formulate development priorities of respective regions. There are also proposals for the establishment of the second tier of self-government with about 308 districts (*powiats*). Figure 9.3 shows new regions approved in Summer 1998.

The specific case of local government exists in the capital Warszawa. At the beginning of the 1990s, the city of Warszawa was a mandatory

association of seven municipalities and the city council consisted of representatives from district councils. Consequently, the city government was the
subject of the individual interests of districts. New administrative division
came into effect in 1994. A single municipality of central Warszawa,
similar to pre-war territory of the city, was created from the former central
district and inner city parts of outer districts. The remaining suburban parts
of former districts were divided into ten relatively homogeneous municipalities. The central city and suburban municipalities form a mandatory
Union of Warszawa (population 1.6 million). Both the municipal and union
councils are directly elected. The Mayor of Central Warszawa is also the
Lord Mayor of the Warszawa Union. The role of the Union is to supervise
metropolitan development and it is in charge of spatial planning, development strategies, infrastructure investments and is in possession of instruments of income equalisation (Surazska, 1996). The Union's income is
independent of municipalities and comes from a share in corporate taxes

Figure 9.3 New regions (from 1.1.2000) and former voivodships of Poland

and fees. Individual municipalities have autonomous property rights. However, the central government has the right to divest Warszawa municipalities, without any compensation, of land and buildings necessary for central government functions, including international organisations (Surazska, 1996).

Regional policies

The most important governmental agency in Poland that formulates and implements regional planning and regional policies is the Department of Physical and Long-Term Planning of the Central Planning Office, which is supposed to formulate perspective economic and physical plans for Poland and to establish foundations for state regional policy (Gorzelak, 1996). In the early 1990s, state regional policy was shaped by the pressure exerted by trade unions in regions with a concentration of negative social effects of economic restructuring. The policy granted subsidies for infrastructure development in old industrial regions most endangered by structural unemployment. The funds allocated in 1991–93, however, were negligible constituting less than 0.2 of the central government spending (Gorzelak, 1996).

The Ministry of Labour and Social Policy is also strongly involved in local and regional intervention and has probably the most developed concept of explicit regional policy focused on regions with a high level of unemployment. Its employment policy delimits areas of high structural unemployment in which economic instruments are used in collaboration with the Ministry of Finance. In 1993, the areas included 412 municipalities accounting for 15 per cent of Poland's population and 20 per cent of unemployed and, recently, further municipalities with rapidly growing unemployment have been added (Gorzelak, 1996). The measures used in these areas comprise accelerated amortisation rates of fixed assets, infrastructure grants for local budgets, income tax relief for private businesses which run vocational training, exemption of firms from income and salary taxes for twelve months (in the case of employing school leavers recruited through Employment Offices), the possibility of firms with foreign capital to apply for income tax relief, and grants from the Work Fund for active forms of coping with unemployment (based on Gorzelak, 1996: 134). There are no official evaluations of these programmes (Gorzelak, 1996) and there is an opinion that 'regional measures applied under the active labour market policy have not worked so far' (UNDP, 1996: 25). Despite the overall unemployment figure falling in 1995, the regional disparities increased, with the highest figures in rural areas.

Regional policies of the early 1990s were characterised by low activity due to the priority given to macro-economic policy, unclear institutional responsibilities and little co-ordination between various governmental ministries and other agencies and very limited funding. It was based on a reactive approach without any attempts to formulate longer-term regional development strategies. In the late 1990s, an important impact on the formulation of a new regional policy came from the association agreement with the European Union.

Policies of regions

Regional authorities are a part of the state administration and they do not conduct their own policies. However, they influence the development of regions by claiming funds and assistance from central government and by helping to organise, create and fund regional development agencies, regional councils, foundations for regional restructuring, etc. (Gorzelak, 1996). The municipalities have their representative assembly (*Sejmik*) in each voivodship.

Regional development agencies in particular (there were over 50 in 1994) are new active actors in regional development. They are created by the Industrial Development Agency in co-operation with the regional administration and local authorities and with the support from chambers of commerce and industry, local firms, banks and business associations, etc. The state represented by the Industrial Development Agency usually contributes to the initial capital, but the agencies should be self-supporting. They should be involved in the preparation of local/regional development strategies, but are rather involved in consultancy services for local firms. In some cases they are involved in implementation of programmes within the PHARE framework.

It is expected that the reform of territorial administration will reinforce the powers of regions[1] and with the introduction of elected authorities will enable them to conduct their own regional development policies (Gorzelak, 1996). Paul (1995) sees the contemporary non-existence of self-government on the regional or district level as one of the major obstacles for regional development.

Local (municipal) development practices

There are 2459 municipalities in Poland and they are in general larger than municipalities in the Czech Republic, Hungary or Slovakia. There is no municipality with less than 1000 inhabitants, while in the Czech Republic about 80 per cent of municipalities have less than 1000 people. The main task of local government is the provision of municipal services (local roads, transport, disposal collection, etc.), education, health and welfare. Local authorities act as an investor in local transport and technical infrastructure. The self-governed municipalities also become owners of former state properties, namely land and housing, which they can sell or lease (for examples of the title transfer from the state to municipalities (see Strong *et al.*, 1996: 211–2). Furthermore, they are of crucial importance for physical planning, regulation of development process and environmental protection. For the sake of co-operation in the field of municipal economy, environmental tasks, etc. over fifty inter-municipal associations have been established.

The power of municipalities is limited by financial constraints. Municipal government expenditure accounted in 1993 for only 12.3 per cent of total government expenditure, which is less than in the Czech Republic and developed countries of Western and Northern Europe (Surazska and Blažek, 1996). More than two-thirds of municipal revenue comes from municipal income and from a share of central taxation. The general grant (18.8 per cent of municipal revenues in 1993, source: Su-razska and Blažek, 1996) is provided by the Ministry of Finance according to a formula based on the population size of municipalities, with large towns receiving a higher grant per inhabitant than small municipalities.

The economic activities of municipal self-government are restricted by limiting municipal borrowing to 15 per cent of the annual budget and by forbidding engagement in economic activities that are not directly related to the delivery of public services (Surazska and Blažek, 1996). In comparison with the Czech Republic, Polish municipalities have lower revenues and capital expenditures per capita and overall are more constrained in their local economic development activities. Furthermore, due to unclear legislation there are conflicts between regions (voivodships) as representatives of the state, and municipalities (*gminas*) as representatives of local interests. Limited skills and pressure of everyday matters is a further reason for low spending, beside the low involvement of municipalities in local economic development (Gorzelak, 1996).

The city of Kraków is an example of a municipality with a clearly defined development strategy. The basic planning document is the Plan of the Development of the City of Kraków (UMK, 1997). It is a five-year plan, which is annually updated. It consists of three parts. First, there is a five-year Plan of Social and Economic Development of the City of Kraków that specifies priorities in several fields, such as health and safety, transport, infrastructure, services and trade, spatial management and conservation, etc. The second document is an annual Economic Programme with detailed specification of priorities for a given year in transport infrastructure, housing, etc. The third document is the five-year Programme of Finance and Investment. This is considered to be the most important and elaborates in great detail all municipal expenditures. In 1998, the municipality was preparing a study of use that was intended to aid preparation of detailed plans for amended spatial arrangements (see the next section on physical planning).

Physical planning and the control of development process

The Communist spatial planning system was oriented to the physical realisation of goals which were contained in national and regional economic development plans. Physical planning was subordinated to economic planning. The legislative background for the system of physical plans was settled in the 1961 Physical Planning Act. This Act together with the 1984 Act on Spatial Planning and various building and environmental laws formed the legal basis of planning during Communism and in the first half of the 1990s (Judge, 1995). The emphasis of the Communist planning system was on the preparation of hierarchically organised long-term regional (voivodship) and detailed municipal plans based on rigid land-use allocation. There were two types of plans used in urban areas, a general city land-use plan, with a strong emphasis on the physical arrangement of the city, and detailed plans used for the regulation of the development process.

At present, the Ministry of Spatial Planning and Construction is responsible for the general building and physical planning rules and for other regulations concerning development process on the local level. A new Building Code was approved by the Parliament in 1993 and a new system of regional and physical planning, based on the Spatial Planning Act (*Ustawa o zagospodarowaniu przestrzennym*) from 1994, has been introduced since the beginning of 1995. The new system of spatial planning defines two basic levels of spatial planning and corresponding actors, the state and municipalities.

The state is involved in spatial planning on the national and regional levels. The Central Planning Office is supposed to formulate the concept of the national plan of spatial arrangement. This document is legally binding only for central government institutions whose policies and programmes have explicit regional targets. On the regional level, the old voivodship plans lost their validity and are replaced by two new documents: the study of spatial arrangement and the regional development programme. These documents are summaries of the state activities in a given region and can also include development goals of regional government. They are not legally binding and have an information and advisory role. The projects incorporated in the study and programme are negotiated with the municipalities. If agreement is achieved and projects from the regional plan are included in the local physical plan it gains a status of legally binding component of planning. The cases where agreement is not achieved between regional government and municipal self-government are decided by the Council of Ministers.

Local physical planning at a municipal level is considered to be the basis of the planning system and only local physical plans are legally binding documents. There are two consequent steps in local physical planning. First, a study of spatial arrangement must be elaborated. It covers all municipal territory, has the form of a general land-use plan and is not a legally binding document. Second, legally binding local plans of spatial arrangement are prepared for parts of a municipal area and have the form of detailed regulation plans. Local plans also include a prognosis of the environmental impact of planned projects. It is not obligatory for the municipality to prepare the new plan. However, in certain cases defined by the law, for instance when there is a project of national interest located on municipal territory, the municipality is obliged to prepare the plan. If municipality does not make the plan in such a case, it will be prepared and approved by the regional (voivodship) government. It is generally expected that old physical plans will be replaced by new ones by the year 2000.

Polish planners often criticise the inadequate regional planning framework. It serves only to facilitate transfer of national development goals defined by individual sectoral ministries to local plans. It leads to a strengthening of centralised sectoral planning over regional planning. The passivity of regional planning can be overcome only with the introduction of regional self-government.

In Warszawa, the old plan from the 1980s was considered too rigid, detailed and outdated. The new Master Plan for Warszawa, that is more suited to market conditions, was approved under the old legislation in 1992. It divides the city into broad zones that define dominant land-use types. The plan for each land-use zone indicates a series of preferences, allowances and exclusions. The main functions of the plan are the co-ordination between local plans of the communities within Warszawa area and environmental protection. It also includes public investment programmes for transport and public infrastructure and public facilities, such as schools or hospitals.

The main regulations concerning the development process are decisions concerning the terms and conditions for construction and land-use in respect of building and planning permits. These decisions must be secured for most developments (they are defined in the Building Law). In relation to a specific site, they determine the development type, terms and conditions resulting from designations contained in the local plan of spatial arrangement (local land-use plan), terms and conditions arising from other regulations and the time period for which the decision remains valid (usually two years). The procedure of issuing the decision takes a maximum of two months from the submission of a complete and appropriately prepared application. Nevertheless, foreign commentators see the granting of planning permission as a bureaucratic and time-consuming procedure which can take as much as fifteen to eighteen months (Judge, 1995).

Building permits are administrative decisions which entitle the recipient to commence construction work. The building permit can be granted only to those who have been granted a valid decision on construction and land-use and who can prove the right to build on the property in question. The application for a building permit must include building plans and all required opinions, approvals and permits stipulated in relevant regulations. The detailed scope and form of the building plan is described in a decree of the Ministry of Spatial Planning and Construction issued in 1994. For structures whose use may pose an environmental hazard, the building permit must include a specialist assessment prepared by a specialised person or organisation designated for this purpose. The processing period takes a maximum of two months. Building permits lose their validity if the construction work does not commence within two years from the date of issuing the permit.

Note

1 On the 26th of July 1998, the Polish parliament (*Sejm*) approved a compromise variant of new territorial division of Poland into 16 regions (voivodships, *województw*) and 308 districts (*powiats*). The population size of the new regions ranges from one to five million inhabitants. The reform brings a radical decentralisation of political power from the central state to regional governments. Regions will be governed by elected regional assemblies and the state administration at this level will be represented by an appointed governor. Regional self-government will play an important role in education, health care, social services and, importantly, in the implementation of regional planning and regional development policies. Regions will become operational on 1 January 1999.

10 Infrastructure, the environment and regional development

Infrastructure

The development of transport, telecommunications and energy infrastructures have a considerable impact on rates of regional economic growth throughout the EU, but, until the early 1990s infrastructural development was undertaken with little reference to the need to reduce regional imbalance in the Community as a whole (CEC, 1991). In response to the Treaty of Maastricht of 1991, the White Paper, *Growth, Competitiveness and Employment* (CEC, 1993b) proposed that, as a priority, trans-European networks (TENs) should be created not only for transport but also for telecommunications and energy to facilitate multi-mode connections with peripheral regions and to help bring about a truly single market (CEC, 1994c).

The White Paper estimated that the development of TENs would require an investment of up to 400 billion ECU over the period 1994–99, but the public sector would be unlikely to contribute more than a quarter of this sum (see Table 10.1), with the private sector being relied upon to provide

Table 10.1 Community funding of TENs, 1994–99

Source	billion ECU
Community budget:	26.5
TENs	2.50
Structural Funds [1]	9.75
Cohesion Fund	11.50
R&D	2.75
EIB (loans)	33.5
Union Bonds (esp. transport & energy)[2]	35.0
Convertibles guaranteed by EIF (esp. telecoms)	25.0
TOTAL	100.0

Source: CEC (1994a)

Notes
1 Some 30 per cent of the Structural Fund is spent on infrastructure investment (most being derived from the ERDF)
2 The issue of Union Bonds would provide the means by which the governments of Member States would contribute to the funding of TENs

the remainder. From the 400 billion ECU, it was intended that 220 billion would be invested in transport, 220 billion on telecommunications and 13 billion on energy.

Transport

In the 1970s and 1980s, the EC was increasingly faced with the growth of traffic, severe congestion and decreasing accessibility – particularly in Greater London, the Randstad, the Ile de France, München, Milano and Madrid. This was a reflection of the failure of investment to keep pace with the growth of both freight and passenger traffic. Whereas between 1975 and 1984 the volume of traffic in the EC increased by 25 per cent, investment declined by 22 per cent (CEC, 1991b). Increased investment, however, was not only necessary to reduce congestion in the central regions but also to improve the infrastructure of the peripheral regions and to enhance connections between the periphery and the centre.

Although, under the Treaty of Rome, transport policy was intended to 'play a vital role in the creation of a common market' (Williams, 1996: 167), the neglect of transport investment in the peripheral regions during the early years of European integration was, in large measure, a result of the European Parliament and the Commission being pre-occupied with such matters as the establishment of the Common Agricultural Policy and the European Coal and Steel Community's programme for the coal and steel sectors (Williams, 1996). An inadequate volume of transport investment was also attributable to the comparatively low levels of peripheral demand, a partial dependence on the transport systems of neighbouring countries, the variable willingness of governments to invest in cross-border routes, and the existence of natural barriers such as the sea or mountain ranges.

In a belated attempt to compensate for this deficiency, most of the 3.5 billion ECU of Community support for road investment in the period 1989–93 was heavily concentrated in the peripheral Member States of Spain, Portugal, Ireland and to a lesser extent Greece. Similarly, in the same period, most of the 1 billion ECU of Community support for rail investment was allocated to projects in Spain, Portugal, Greece and Northern Ireland, while 500 million ECU was invested in ports and airports mainly in the peripheral regions (CEC, 1991b).

Notwithstanding these developments, the transport infrastructure had not been a central cause for concern among the transport ministers of the Community. As a direct consequence of the *Gendebien Report* (EP, 1983) – which highlighted the lack of action in transport planning – the Council of Ministers were ruled negligent by the European Court of Justice for failing to produce legislation for a common transport policy. As an outcome of the Single European Act of 1986, a Committee of Transport Infrastructure was set up to help hasten the introduction of the Single European Market, but the formulation of an EU transport strategy had to await a new sense of urgency imposed upon the sector by the Maastricht Treaty of 1991 (Williams, 1995).

Following the signing of the 1991 Treaty, and the establishment of the Single European Market on 31 December 1992, the Member States of the EU at last recognised the urgency of extending and modernising the transport infrastructure of much of the Continent. Set up by the European Council in 1993 (and comprising representatives of Heads of State and government), the Christophersen Group submitted a short-list of priority projects to the Corfu Summit in June 1994. Of a total of thirty-four projects, eleven were to start by 1996 at the latest, a further ten would be accelerated to start in 1996 or as soon after as possible, and thirteen would be subject to further study (CEC, 1994a: 56–7). However, to take account of the potential enlargement of the EU, the number of transport projects scheduled to start by 1996 was increased to fourteen at the sub-sequent Essen Summit of December 1994 (see Figure 10.1). A start would thus be made on developing large trans-European transport networks to help create more spatially balanced economic activity in Europe 'by increasing the potential competitiveness of peripheral regions through an improvement in their accessibility relative to more central areas' (CEC, 1994a: 63). Transport TENs, in the view of the European Commissioner for Transport, the Rt. Hon. Neil Kinnock, would 'not only help reduce transport times and costs: the investment involved will create new business as well as stimulating research and innovation in new technologies, and the creation and operation of some infrastructures will also provide the basis for permanent new enterprises and, most importantly, new jobs' (Kinnock, 1998: 73).

According to the 1993 White Paper, the development of transport TENs should satisfy the following: the need for faster, safer and cheaper transport to facilitate competition both within the EU and externally; the need to create a greater spatial balance in the distribution of population and wealth; the need to seek the optimal combination of different modes of transport to improve performance and reduce environmental affliction; and the need to establish links with countries in Central and Eastern Europe and in the Mediterranean for the stimulation of trade and the development of economic partnership (CEC, 1993c).

Undoubtedly, transport TENs required the investment of very considerable sums of money. The White Paper estimated that total investment would amount to 220 billion ECU over the period 1994–99, but most of this was provided by the private sector, since about only a quarter emanated from Member States and Community sources due to budgetary constraints. In its role as facilitator and co-ordinator, the EU was only able to use its TENs budget (equivalent to less than 1 per cent of the total cost of investment) to fund initial feasibility studies (although the EU has met up to 10 per cent of the cost of any one project of common concern). In contrast to the Structural Fund and Cohesion Fund, the TENs budget contributed only to programmes that would have been viable if funded

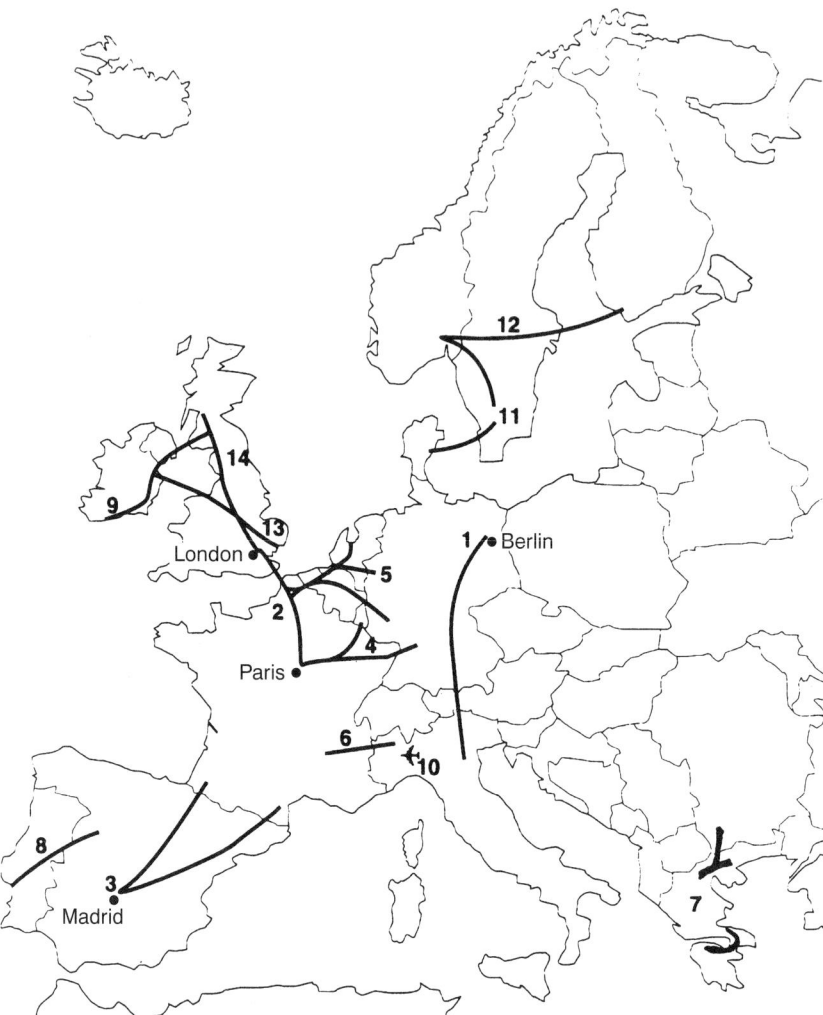

1. TGV/combined transport: Verona-Berlin
2. TGV: Paris-Brussels-Cologne-Amsterdam-London
3. TGV: Madrid-Barcelona-Perpignan; Madrid-Victoria-Dax
4. TGV: Paris-Metz-Strasbourg; Paris-Metz-Saarbrucken-Mannhein;
           Paris-Metz-Luxembourg
5. Conventional rail/combined transport: Rotterdam-German border (Rhein-Ruhr)
6. TGV/combined transport: Lyons-Torino
7. Motorway: Greece-Bulgaria (jointly with East-West motorway project)
8. Motorway: Lisboa-Valladolid
9. Rail link: Cork-Dublin-Belfast-Larne-Stranraer
10. Airport: Malpensa (Milano)
11. Fixed rail/road link: Denmark-Sweden
12. Road-rail/combined transport: Norway-Sweden-Finland
13. Road/combined: Ireland-UK-Benelux
14. Rail: UK west-coast main line

Figure 10.1 Priority projects, transport TENs, EU
Source: Williams, R.M. (1996)

exclusively by national private-public partnerships, but projects that were capable of being funded solely by the private sector were ineligible for support from the TENs budget (Williams, 1996). However, in the four Cohesive Fund countries (Spain, Portugal, Greece and Ireland), the EU contributed up to 80 per cent of the capital cost of eligible transport projects in the period 1994–99, and, since the Edinburgh Summit of 1992, finance and loan guarantees for transport projects have been available in turn from the European Investment Bank and the European Investment Fund.

In the early twenty-first century, the development of transport TENs will undoubtedly increase accessibility in absolute terms and thereby enhance the competitiveness of the EU in world markets, its effects on the regions of the EU are uncertain. On the one hand there is a danger that, as a result of improved transportation, economic activity and population will increasingly concentrate in the areas of maximum accessibility, that is in Belgium, the Netherlands, Luxembourg, northern, central and eastern France, most of Germany, south-east England and northern Italy. If this occurs, the relative accessibility and economic attraction of the more peripheral areas will decline, particularly those which are some distance from TGV stations, motorways or large airports (CEC, 1994c). On the other hand, it is argued that as a result of improved accessibility 'there are likely to be significant gains in Greece, Ireland, the southern and western regions of the Iberian peninsula, Mecklenburg-Vorpommern and Nord Pas-de-Calais . . . [and that plans to develop transport TENs] appear to reduce the degree of peripherality of outlying regions and therefore open up new markets to producers located there' (CEC, 1994c). Similarly, although it is clear that business travellers from cities in the heart of Europe, such as Paris, London, Brussels, Frankfurt-am-Main, Stuttgart, Munchen and Milano can travel in the least time to destinations across Europe, the more peripheral cities of Glasgow, København, Berlin, Rome, Madrid and Athinai are not significantly disadvantaged in respect of air travel to European and world-wide destinations. Investment in peripheral airports will thus not only help to consolidate the European air transport network but will stimulate further economic development based on the export of high value/low weight products to EU or world-wide markets (CEC, 1994c).

The Channel Tunnel

Notwithstanding the anticipated benefits of this investment, it has been predicted that the development of high-speed means of transport between major centres would impose losses on certain regions as well as confer benefits on others. According to a Commission study (ATC Consultants, 1991), the regional economies benefiting most from the Channel Tunnel and the associated development of high speed rail services are likely to be concentrated in regions close to the tunnel (notably in Kent, Nord-Pas-de-Calais, West Vlaanderen and Hainault) and, with the development of the high-speed network, eventually diffused across much of France, Belgium, western Germany and north-west Italy. In contrast, economic 'grey' zones might emerge in, for example Normandie and Zeeland, while Ireland, Northern Ireland, northern Scotland, Denmark, east Germany and most regions in southern Europe are likely to be adversely affected. The study suggested that this could be mitigated by the strengthening of transport links between the peripheral regions and the centre through connections to the high-speed network and the modernisation of ports.

Central and Eastern Europe (CEE)

The EU (through its PHARE and TACIS programmes), the UNECE and the Council of Europe's European Conference of Ministers of Transport perform an important role in helping to establish transport as a key element in the economic transformation of CEE countries. Ministers, at their second pan-European conference in Crete in 1994, identified a set of nine priority transport corridors in CEE (see Figure 10.2), each of which would have road and rail components, be economically viable, be capable of being connected with the EU's transport TENs programme, be able to link the EU regions with the regions of third countries, and facilitate international flows (Williams, 1996).

Information technology and telecommunications

Potentially, new developments in the field of information technology and telecommunications (ITT) could confer substantial benefits on the process of spatial integration within the EU. Information could flow instantly within and between firms regardless of location, thereby improving the economy of even the most peripheral regions (CEC, 1991b). Throughout much of the 1990s, however, there were considerable regional variations in access to modern telecommunication networks. Networks were concentrated in the large markets at the centre of the EU, while studies showed 'that peripheral regions in the south of Europe in particular suffered from the poor quality of even very basic telecommunication services' (CEC, 1991b: 94).

The absence of large markets for ITT services in peripheral areas had rendered investment in extended telecommunication networks unprofitable. Peripheral areas were disadvantaged by a deficiency of transport links to and from the centre, which limited the opening up of commercial markets in general and ITT markets in particular.

Unequal access to information clearly provides centrally-located firms with a competitive advantage over firms located elsewhere. At the end of the twentieth century, a key question was whether the further development of ITT, and its spatial extension, would increase or decrease regional economic disparities. Would, for example, an extension of ITT encourage economic activity to be increasingly located in the centre, or relocated from

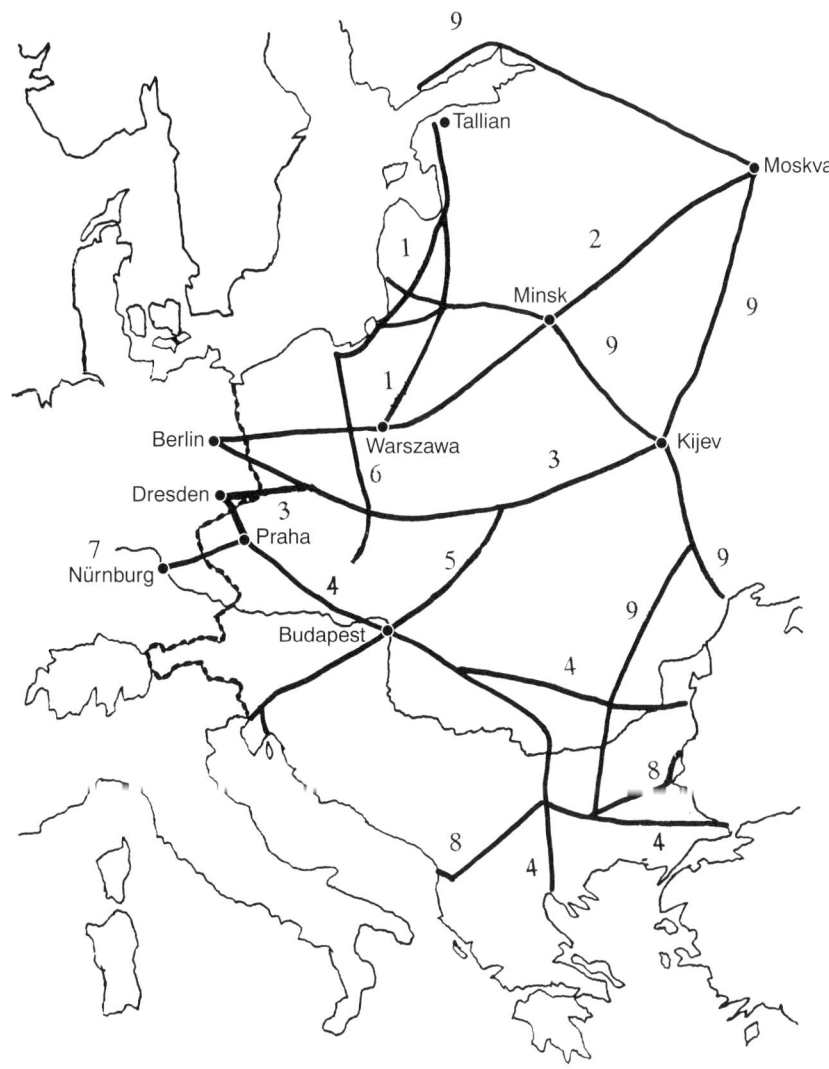

Figure 10.2 Priority transport corridors, Central and Eastern Europe
(Derived from: Cole and Cole, 1997, p. 333)

- - - - - - EU limit (1994)

1.   Tallinn-Warszawa; Riga-Gdansk
2.   Berlin-Moskva
3.   Berlin-Kijev; Dresden-Wroctaw
4.   Dresden-Thessaloniki;
     Sofija-Istanbul; Nürnberg-Constanta
5.   Trieste-Lvov;
6.   Gdansk-Zilina;
7.   River Danube ports in CEE countries
8.   Durres-Varna
9.   Plovdiv-Helsinki; Odessa-Klaipeda;
     Kijev-Moskva

the centre to the periphery. In the view of the Commission in 1991, the development of ITT in the large firm sector could 'perpetuate the concentration of decision-making in central areas while facilitating decentralisation of lower level activities to the periphery'; while in the small firm sector (by facilitating the extended use of micro-processors and the adoption of flexible production systems), ITT enables firms 'to locate away from established high-cost centres without suffering a significant competitive disadvantage' (CEC, 1991b: 91).

In 1994, in more unequivocal terms, the Commission held the view that the development of TENs in the form of telecommunication projects would, in general 'reduce regional disparities in economic activity and employment and enable the less developed areas to take advantage of the possibilities created by the "new European information society" which is in the process of being constructed' (CEC, 1994c). To this end, the Bangemann Group (set up by the European Council in 1993) identified ten priority programmes which would have a major effect on spatial development: teleworking; distance learning; telematic services for small and medium-sized enterprises; road traffic management; air traffic control; electronic tendering; home connections for electronic newspapers; information exchange networks among universities and research centres; health care and other public authority responsibilities; and urban information highways. Although it was estimated that the total investment required would amount to 150 billion ECU over the period 1994–99, Bangemann (1994) proposed that, since market forces would ensure that the extension of telecommunications would be profitable, investment could be left to the private sector – with the Commission limiting its role to that of facilitator and co-ordinator.

Energy supply

Since a secure supply of energy at competitive prices is an essential pre-requisite for the consolidation of a single market, in the late 1980s the Commission aimed to promote the strengthening and further integration of the distribution networks for electricity and gas, and to develop new electricity generating capacity. The provision of new capacity and the extension of networks, however, had to 'be consistent with the aim of promoting the economic development of the weaker regions while remaining sensitive to the needs of the environment' (CEC, 1991b: 97).

There were still a number of major gaps in the system of energy networks. Gas networks did not exist in Greece and Portugal, in the case of natural gas Spain and the UK were not connected to the Community network, while Ireland remained entirely unconnected to any of the networks which were in place. However, previously separate national markets in energy began to merge as electricity and gas networks were conceived and implemented on a Community basis. The Commission was keen to ensure that all regions benefited from a single market in energy and urged, first, 'the adaptation and integration of networks across the Community as a whole'; and, second, better management of energy demand at the regional and local level and the assessment 'of the role of energy in the overall economic development of a region' (CEC, 1991b: 101).

Whilst intra-community imports of electricity increased four-fold between 1950 and 1990, these still accounted for only 8 per cent of total consumption in 1990. Since the total cost of supply could be reduced substantially if the system of interconnections were complete (a saving of 10 per cent or 3 billion ECU at 1990 prices being predicted), there was every incentive for increasing the extent of cross-border transmission (CEC, 1991). The Commission thus saw the need to strengthen the following inter-country links:

- in the north of Europe, links between: Ireland and the UK; France, Belgium and the Netherlands; France and Germany; Germany, Denmark and Sweden
- in the south of Europe, links between: France and Spain; Spain and Portugal; France and Italy; Italy and Greece; Italy, Switzerland, Austria and (former) Yugoslavia.

In addition, it was necessary to improve the wholly inadequate links between the old West Germany and the five *Länder* of former East Germany, as well as between the EU and CEE.

The extension of natural gas transmission networks was also considered essential for the consolidation of a single market and regional development. The demand for natural gas (projected to increase by about 50 per cent between 1990 and 2010) would not be satisfied unless all parts of the EU were connected to the network (CEC, 1991b). In the short term, in Northern Europe, the Commission decided to promote the extension of pipelines from the UK to Ireland and the connection of western Germany with the new *Länder*, while in Southern Europe gas infrastructures would be developed in Greece and Portugal, and there would be an increase in the capacity of the trans-Mediterranean pipeline and the construction of a pipeline between Algeria and Spain, and Spain and France. In the longer term, the Commission foresaw the completion of networks with the UK being linked to the Continent, Scandinavia being linked to the rest of Europe, and North Africa being more fully linked to Spain, France and Italy (CEC, 1991b). In the early 1990s, the REGEN initiative accelerated the introduction of transmission networks in some of the peripheral regions, notably Greece, Ireland, Portugal, Corse and Sardegna – and their inter-connection with Europe-wide networks (CEC, 1991b). In this first phase of the programme, the Community co-financed four projects to the sum of 347 million ECU.

As far as new electricity generating capacity was concerned, the Commission recognised that the construction of generating plants on new sites would be unavoidable but, rather than promote the further development of currently available sites, the Commission declared a preference for the construction of small (mainly gas-powered) decentralised units suited to local needs and the protection of the environment.

Imports of natural gas from non-member countries, particularly Norway, the USSR and Algeria, were projected to increase from 38 per cent of Community consumption in 1990 to 50 per cent in 2000 – necessitating technical co-operation between Member States and their suppliers, and further co-operation among the Member States themselves with regard to the development and integration of transmission networks (CEC, 1991b).

Community policy on energy networks continued to aim at increasing the security of energy supply in relation to demand, improving competitiveness and strengthening economic and social cohesion. While the production of energy in the EU was projected to decline by around 30 per cent up to 2020, total demand for energy was expected to increase by 1 per cent per annum to at least 2010. The Union will thus become more and more dependent on supplies of energy from (as yet) non-member countries – requiring the construction of an increasing number of transnational supply lines (high tension electricity lines and oil and gas pipelines) and the further development of sources in CEE and North Africa.

Competitiveness will be enhanced if consumers are allowed to exercise choice between different forms of energy. To this end, the extension of electricity and gas networks will not only assist regional development, but also facilitates choice, while the liberalisation of energy markets – as proposed by the Commission – should provide an incentive for energy suppliers to improve their competitiveness.

Within the EU, economic cohesion is not only dependent, in part, on the availability of energy, but on the ability of consumers to exercise the same degree of choice regardless of location. In the mid-1990s this was far from the case, particularly in the peripheral regions of Scotland, Ireland, Portugal, western Spain and Greece. Following a report produced by the Christopherson Group, the European Council decided at the Corfu Summit in June 1994 to approve the continuation of feasibility studies on a priority list of eight energy TENs (Table 10.2). Undoubtedly, the development of energy TENs would be costly. The 1993 White Paper had estimated that the total cost of investment would amount to 13 billion ECU over the period 1994–99. Since it could not be expected that Member States would contribute more than one-third of this sum, and the EU more than 1 per cent (or up to 10 per cent for individual projects of common interest), financing would need to be provided by public-private partnerships (Williams, 1996). However, up to 85 per cent of the capital cost of projects in Greece, Ireland, Portugal and Spain could be met by the Cohesion Fund, while additional finance could become available from the European Investment Fund and EIB.

Two further energy TENs were approved at the subsequent Essen Summit. Both were natural gas pipelines, the first connecting Algeria and

Table 10.2 Priority energy projects, 1994–99

1. Italy-Greece electrical connection (cable)
2. France-Italy electrical connection
3. France-Spain electrical connection
4. Spain-Portugal electrical connection
5. Electrical connection between the western and eastern parts of Denmark (cable)
6. Introduction of natural gas in Greece
7. Introduction of natural gas in Portugal
8. Connection of natural gas networks of Portugal and Spain

Source: CEC (1994a)

Morocco with Spain, and the second linking Russia, Belarus and Poland with Germany.

In addition, the Commission continued to consider some sixty energy projects designed to improve supply to peripheral regions, while the REGEN programme will ensure that some of the missing links in the trans-European energy networks will be completed, notably by means of a natural gas pipeline between the UK and Ireland, an undersea cable between Greece and Italy, and gas distribution networks in Greece and Portugal.

Infrastructure and the Single European Market

Clearly, the development of transport, telecommunication and energy TENs is central to the realisation of the Single European Market since the free movement of goods and services is manifestly dependent upon efficient, competitive, safe and environmentally-friendly infrastructural services. According to the Commissioner for Transport: 'If we are really serious about building a wider Europe, integrating our economies and increasing competitiveness, the dislocations that come from inadequate infrastructure and fragmented administration must be overcome' (Kinnock, 1998: 73).

By the end of the 1990s, it was too early to judge whether or not the TENs had been successful in generating economic development and economic cohesion, but it is probable that – outside of the Cohesive Fund countries – an overwhelming reliance on private sector resources might have prioritised investment in the central areas of the EU rather than along the routes that link the many peripheral or economically-backward regions with the existing principal markets of the Union (Williams, 1996).

It was also premature to assess the extent to which regional and local authorities throughout the EU (and not least in cross-border areas) had been able to incorporate a consideration of TENs and secondary networks into their regional and local planning processes. However, if smaller towns and rural communities are inadequately linked to new infrastructural development, spatial disparities of economic and population growth will widen and larger regional cities will prosper at the expense of their smaller neighbours (Williams, 1996).

Regardless of whether or not extended infrastructures have the potential to play a major role in the inter-regional or intra-regional polarisation of economic activity, environmental considerations have become increasingly central stage in the development process throughout the EU. Although, by the late 1990s, the Council of Ministers had not agreed on an acceptable form of strategic environmental assessment to identify the environmental attributes of TENs and their secondary networks, it can only be a matter of time before the decisions taken at the Corfu and Essen Summits in 1994 (to accelerate and facilitate the creation and development of networks) will need to be modified on environmental grounds in an attempt to add an element of sustainability to the concept of the Single European Market.

The environment

Interest in the environment has been growing within Member States and the European Union, and indeed around the world, over the last two decades. The driving forces have been the developing scientific consensus that there is a potential for environmental catastrophe and that human impacts are a significant part of the impending crisis. It is also clear that individual actions are not going to be sufficient to provide a realistic answer and that collective responses by governments, citizens, and international organisations will be necessary.

The EU has had a long-standing involvement in environmental policy, even though the environment was not directly incorporated into the original treaty. It was not until 1986 under the Single European Act that a specific requirement was included to look at environmental matters. However, from the 1960s various environmental directives and regulations have been implemented and from 1973 onwards the EU has undertaken a series of environmental action programmes which have structured more broadly based policy and action. Essentially, environmental responses have become accepted as part of general EU policy and implementation processes, to the extent that Vogel could maintain, even before more recent enhancements, that 'it is clear that at least in the area of environmental regulation, the Community is already a federal structure'. (Vogel, 1995: 129).

The reasons for Community action in the environmental field, despite what may be perceived as a weak legitimacy at the beginning, can be seen in a variety of aspects: public concern combined with increasing political pressure to strengthen the need for environmental action. In a 1973 public opinion survey environmental pollution was seen as the most important problem in the EU. By the end of the 1980s public opinion manifested itself in political life with the European Green Parties having thirty-seven seats in the European Parliament. There was also, in the 1970s and early 1980s, a role for environmental policy and action in preserving the momentum of European integration which, in other contexts, had stagnated. The development of environmental responses has an economic justification in terms of the competition or single market argument. The need for a basic level of environmental regulation stops any one state acquiring an unfair economic advantage. In other words, the environmental regulations ensure a level playing field for all competitors. More recently, the economic arguments have been boosted by the development of more positive economic benefits, arising from the creation of particular sectors or industries based on environmental factors. So a clean environment may be beneficial to tourism or to more general business attracted by high quality areas. Moreover, some business is developing from the environmental industry itself. Finally, these broad reasons, underpinning interest in environmental policy and programmes, were related to the geography of Europe in that environmental pollution does not stop at regional or national boundaries. Concern with river pollution was an early aspect of EU action (Vogel, 1995; Williams 1996).

Prior to 1986, the basis for Community intervention was the general provision of Article 100 of the 1957 Treaty of Rome which was basically concerned with the harmonisation of individual state law and action in order that the Common Market might not be harmed. This was supplemented by Article 235 which permitted action in fields which were not otherwise covered by specific fiat within the treaty (Fairclough, 1983).

This rather weak basis was eventually strengthened by the Single European Act 1986. Article 100A enacted environmental quality as a legimate Community objective in its own right, for the first time. It, further, required environmental protection to be seen as a component of the other policies of the Community. The environmental protection to be sought should be at a high level, though national standards could be higher than the standards set by the Community. It also became possible for directives formed under Article 100A to be passed by qualified majority voting rather than unanimously, as had been the case before.

The 1986 basis was enhanced through the Treaty on the European Union (the Maastricht Treaty) of 1992. The treaty added to the armoury of Community action to permit the promotion of measures at the international level to deal with regional or global environmental problems (Article 130r) and, for the first time, allowed the Council to adopt measures relating to town planning and land-use (Article 130s). However, it is also notable in specifying the principles under which environmental action should be taken. It states that policy should be 'based on the precautionary principle and on the principles that preventative action should be taken, that environmental damage should as a priority be rectified at source and that the polluter should pay' (Article 130r(2)). A second major principle, of subsidiarity, is contained in Article 3b. This was then elaborated at the Edinburgh Summit in December 1992 (Williams, 1996).

These principles effectively mark the summation of ideas and approaches over a considerable period of time, with aspects drawn from the Community and other sources. In particular, the developing Community environmental policy programmes have provided a background in principles and action. Starting with the first programme in 1973 the current periodic programme is the fifth which runs between 1993 and 2000 and is based on sustainability.The concept of sustainability is the most significant and, at the same time, the most elusive principle to have been developed in the environmental field. Although a variety of definitions have been proposed, the Brundtland Report definition is the most familiar – 'to ensure that it (development) meets the needs of the present without compromising the ability of future generations to meet their own needs' (Brundtland, 1987: 8). The Fifth Environmental Action Programme of the Community used this as a basis for the current period by:

- focusing on sources rather than receptors of pollution;
- identifying five main target sectors of EU relevance: industry, energy, transport, agriculture, tourism;
- addressing behaviour patterns of producers, consumers, citizens and governments;
- increasing the emphasis on shared responsibilities (Nugent, 1994).

The progress report on the fifth programme assessed the influence of the programme on the environment and on the progress towards sustainable development. The conclusion was 'one of cautious optimism', and went on to suggest that while the 'overall strategy and objectives . . . remain valid; what is lacking are the attitude changes and the will to make the quantum leap to make the necessary progress towards sustainability.'(CEC, 1997b: 10).

Within the EU the emphasis on the environment has steadily increased alongside the national and international concerns. There has been a change from what might be termed 'theory' towards a more realistic appreciation of the practicalities of implementation. This has been accompanied by a development of the principles underpinning environmental action and a coupling of these principles with action. The organisations responsible for the environment have been enhanced by the work of the European Environment Agency since 1994, with its focus on ensuring that the information basis of environmental action is made more robust. Knowledge of the actions and effects of environmental policy is a new and developing area and the European activity has been significant in progressing these elements. The realism incorporates a recognition of the difficulties of implementing environmental action and of the need to encourage a greater participation if the required results are to be achieved.

The impacts of environmental policy and action are clearly difficult to assess (Hitchens, 1997), however, there are inevitably varying effects on different sectors with, for example, the agricultural sector showing less integration of environmental considerations than the manufacturing sector

(CEC, 1997b: 11). It can also be argued that particular regions may exhibit varying responses and reactions to environmental action. (Hitchens, 1997), with the richer regions being more able to respond to environmental concerns and more able to benefit from any economic advantages to be derived from new sectors and new products related to the environment.

The review of the fifth environmental action programme notes that a number of factors will impact upon environmental implementation in the future (CEC,1997b). The increasing infiltration of environmental action into those sectors, which until now have been less than active, is one area of development. A second is the widening realisation of more people of the need for action. However, the extension of the EU into new members may create fresh strains. It is reasonable to assume, though, that the environmental emphasis is likely to grow rather than diminish over the foreseeable future and that governments, businesses and individuals will have to adapt to these environmental requirements.

11 Urban areas

Introduction

At the end of the twentieth century, the European Union was the most urbanised region of the world with around 80 per cent of its total population living in towns and cities compared with 77 per cent in Japan, 76 per cent in the US, 67 per cent in Central and Eastern Europe, and 35 per cent in the developing world (CEC, 1991b). In all EU countries, the process of urbanisation had been continual throughout the century, but since the 1960s it had been generally slower in the industrialised and highly urbanised north than in the more rural north or less urbanised south, where the movement of people from the countryside to the towns proceeded at a considerable rate (Table 11.1).

The spatial pattern of urbanisation in the EU is characterised by a close network of towns and cities of varying sizes. In 1990, there were 3560 urban areas of more than 10,000 inhabitants with a total population of 237 million, in contrast to only 1000 of this size in both the US and Japan. Within the EU there were also 169 cities of more than 200,000, of which there were thirty-six of more than a million – accounting for around 56 per cent of the urban population (CEC, 1991b; Population Reference Bureau, 1995). Table 11.2 shows that of the 36 'million plus' cities, fourteen are national capitals, and that, in addition to the six largest urban centres, there are two very large clusters of cities: one in North-Central England (Birmingham, Greater Manchester, Leeds/Bradford, Liverpool and Sheffield), and the other in northern France, Belgium and the Netherlands (Lille, Bruxelles, Rotterdam and Amsterdam) containing populations of respectively 10 and 8 million (Population Reference Bureau, 1995). Eight of the largest concentrations of urban population in the EU, moreover, are located in the relatively small Paris, Frankfurt-am-Main, Manchester triangle. It is of note that within Central Europe (beyond the eastern boundaries of the EU), there are a number of cities with populations of more than one million, for example, Budapest, Praha, Warszawa and Katowice, each of which is closer to the urbanised core of the EU than Athinai, Helsinki and Lisboa (Cole and Cole, 1997).

Table 11.1 Urban population and the increase in urbanisation in the EU, 1960–95

Rank urbanisation		Urban population		Increase in population
		%		%
		1960	1995	1960–1995
1	Belgium	92	97	5.4
2	UK	86	92	7.0
3	Netherlands	85	89	4.7
4	Luxembourg	62	86	38.7
5	Germany	76	85	11.8
6	Denmark	74	85	14.9
7	Sweden	73	83	13.7
8	France	62	74	19.4
9	Italy	59	68	15.3
10	Spain	57	64	12.3
11	Finland	38	64	68.4
12	Greece	43	63	46.5
13	Ireland	46	57	23.9
14	Austria	50	54	8.0
15	Portugal	22	34	45.5

Source: Population Reference Bureau (1995)

The importance of larger cities to the national economies of the EU cannot be overestimated. In total, with populations ranging from 12.7 per cent of the national population in Belgium to 38.7 per cent in the UK (Table 11.2), cities of one million or more are the main source of prosperity in their respective countries and, because of higher productivity, contribute disproportionately more to national GDPs than medium-sized cities (CEC, 1997c). However, GDP growth attributed to large cities has often not been matched by employment growth. Whereas the urban regions of Bruxelles, Rhein-Ruhr and London had annual GDP growth rates of 5 to 6 per cent in the 1990s, annual job-creation over the same period was only + 0.2 per cent in Bruxelles and +0.1 in Rhein-Ruhr, but −0.2 per cent in London.

The urban hierarchy

From an examination of the spatial distribution of towns and cities within the EU (together with associated data), it can be argued that there is no clear urban hierarchy in Europe (Wegener, 1995). Whether one chooses functional indicators or accessibility measures, it is apparent that all gradations of centrality are haphazardly scattered across the continent. Kunzmann and Wegener (1991), however, had previously suggested that a four-level hierarchy of cities of 'European importance' has emerged in recent years. In first rank order are Paris and London, the only two global cities in the EU

Table 11.2 Cities within the EU with populations of at least one million, mid-1990s

Rank		Country	Population (millions)	% National population
2	London*	UK	10.3	
11	Birmingham	"	2.6	
12	Greater Manchester	"	2.6	
13	Leeds/Bradford	"	2.1	
20	Liverpool	"	1.4	38.7
24	Glasgow	"	1.3	
25	Sheffield	"	1.3	
29	Newcastle	"	1.1	
7	Athinai*	Greece	3.5	33.3
15	København*	Denmark	1.7	32.6
35	Dublin*	Ireland	1.0	27.0
22	Lisboa*	Portugal	1.3	26.3
23	Porto		1.3	
14	Wien*	Austria	2.0	24.7
4	Milano	Italy	5.1	
8	Roma*	"	3.1	
10	Napoli	"	2.6	23.2
16	Torino	"	1.6	
30	Genova	"	1.0	
3	Rhein-Ruhr	Germany	9.2	
6	Berlin*	"	4.6	
17	Hamburg	"	1.6	22.0
19	München	"	1.5	
28	Frankfurt/Main	"	1.1	
1	Paris*	France	10.7	
27	Marseille	"	1.1	22.0
31	Lille	"	1.0	
5	Madrid*	Spain	4.9	
9	Barcelona	"	2.7	22.0
32	Valencia	"	1.0	
36	Helsinki*	Finland	1.0	19.6
18	Stockholm*	Sweden	1.5	16.9
33	Rotterdam	Netherlands	1.0	12.9
34	Amsterdam	"	1.0	
21	Bruxelles*	Belgium	1.3	12.7

* National capital
Source: *Population Reference Bureau, 1995*

which, in terms of population, compare with, say, New York, Tokyo, Mexico City, Cairo or Bombay. Of second rank importance are the conurbations of Rhein-Ruhr (Bonn/Köln/Düsseldorf-Dortmund/Essen/Duisburg), Rhein-Main (Frankfurt), København/Malmö, Manchester, Leeds, Liverpool, and the Randstad (Amsterdam/Rotterdam). Of similar European importance are a number of larger cities ('Eurometropoles') such as Athinai, Barcelona, Berlin, Birmingham, Bruxelles, Hamburg, Madrid,

Milano, München, Roma, Wien and Zürich – all of which perform essential economic, financial or political and cultural functions on a European scale. With the enlargement of the EU in the first decade of the twenty-first century, certain cities in East Central Europe would soon be added to this list, such as Budapest, Warszawa and Praha. The third rank of cities comprises some national capitals such as Dublin and Lisboa, and other cities of European importance such as Glasgow, Napoli, Palermo, Strasbourg, Stuttgart and Torino – although their function is more national rather than continental. Kunzmann and Wegener suggest that, below this level, various lower urban hierarchies follow, depending on national definitions of central places (for example, as in Germany, Austria, the Netherlands and Denmark), and that as Europe becomes increasingly integrated these national urban hierarchies will be superseded by a single urban hierarchy throughout the EU.

Within the larger cities of the urban hierarchy, irrespective of their European importance, there is often a variety of problems linked to infrastructure, transport, economic development, land prices, housing and the environment, and considerable disparities in the quality of life of different socio-economic and ethnic groups (CEC, 1991b). However, the spatial dimension of the urban economy and its social manifestations are not static, but dependent upon the processes of urban growth and continual changes in the internal structure of cities.

The internal structure of cities

According to the Stages theory of urban growth (Van den Berg, *et al.*, 1982; Vanhove and Klassen, 1980), as the industrialisation process proceeds urban areas will move slowly from one stage to the next. In the first stage of urbanisation, agricultural labour migrates from the surrounding rural areas to the fast-growing industrial sectors located in the cities. Such changes were evident as early as the late eighteenth century in Britain, but occurred only during the nineteenth century in most of North-West Europe and even later in much of Southern and Central Europe. At this stage, population growth in the urban 'core' is rapid, whereas the population in the surrounding 'ring' may actually decline as resources and population are drawn to the town and city.

In the second stage, extensive transport facilities are created as well as public amenities and better housing. The role of services expands and manufacturing industries are moved farther away from the centre. With better public transport, and particularly, growing car ownership, an increasing proportion of households takes advantage of suburban residence to secure lower density housing in a quieter environment while still maintaining reasonable access to the city. Planning authorities rarely resisted these tendencies and eventually population growth in the surrounding 'ring' comes to exceed that of the core area itself.

By the third stage, the population continues to suburbanise, but at the core the population falls, since residential uses there are increasingly coming into conflict with other uses, especially offices which achieve higher land values. Overall, however, the urban area continues to grow, albeit more gradually. It is also during this stage that the problems caused by excessive suburbanisation become acute; the existing road network in particular can no longer cope with growing numbers of commuters and congestion reduces the accessibility of workplaces near the city centre (Van den Berg, *et al.*, 1982). The noise and nuisance of living in the core provide an added incentive for residents to suburbanise, whilst others were compelled to move when housing was bulldozed to make way for additional road and parking capacities.

During the final stage(s) the preceding effects become more acute and spread also to the suburban area or ring which may in turn decline, while the population of the broader region in which the city is located might expand. As part of this process, satellite towns at a distance of between 50 and 100 km from the 'core' city will themselves be expanding at an earlier stage in the urban life-cycle – thereby encouraging job loss in the original urban system (Van den Berg, *et al.*, 1982).

Within the EU in the 1980s, and at the third and final stage(s) in the urban life-cycle, population increased in the regions in which the few cities of over 2 million were located, but the number of people living in the cities themselves did not always grow by as much or simply declined. While the population continued to expand in the Paris conurbation, Madrid and Lisbon – albeit at a lower rate than in their regions, in Greater London, population declined by 56,200 in the 1980s largely as a result of a net outward migration of 302,000 inhabitants, mainly to areas relatively close by, and in København the population similarly declined due to migration to the surrounding region or further afield – the growth of population in the greater København area being less than in other parts of Denmark (CEC, 1991b).

The Stages theory is probably most useful in stressing the interrelationships between the growth of the urban core and the periphery, and because it points to the cumulative nature of urban decline. However, although the theory recognises the possibility of urban resurgence (a fifth stage), this is normally not assumed to be a consequence of unbridled market forces but rather the result of concerted efforts on the part of central and local governments to encourage urban renewal, housing improvement and to ease the traffic problem (Vanhove and Klassen, 1980).

European urban life-cycles

The pattern of European urban development from 1950 to the millennium clearly illustrates the validity of the Stages theory. Dividing the period to 1975 into three sub-periods, Van den Berg *et al.* (1982) found that, of the countries studied, only in Bulgaria, Hungary and Poland did the dominant stage of urban development fail to progress. All other countries moved up one or two stages and in no case did any country move back a stage. These results, shown in Table 11.3, illustrate, first, the evolutionary nature of the European urban system and, second, that it is not individual cities but the urban system as a whole which is affected by similar structural changes.

Clearly, Table 11.3 fails to include a number of peripheral countries such as Greece, Ireland, Portugal and Spain as well as, for example, the former state of Czechoslovakia in Central Europe. The urban population in these countries grew at a rapid rate throughout the 1960s and 1970s, with urban life-cycles, in consequence, remaining at the urbanisation or 'urbanisation/ suburbanisation' stages. The table also fails to show how the cycle developed during the last quarter of the twentieth century. It is probable that most countries (with the possible exception of those in Central and Eastern Europe) moved up a stage in the urban life-city, whilst in several countries a further stage in the cycle might have emerged – that of resurgence or 're-urbanisation'. In West Germany, this, in part, would have been the result of an inflow of migrants from Central and Eastern Europe as well as from the eastern *Länder*.

Recent data indicate that in the second half of the 1980s urban decline was confined to Italy, Switzerland, Belgium, the UK and parts of France, while some of the older metropolitan areas of the North have experienced a significant economic resurgence (CEC, 1991b). Whereas, for example, routine office functions continued to be decentralised, there was simultaneously

Table 11.3 Changes in the classification of countries by dominant stage of urban development

Dominant stage	1950–70	1960–70	1970–75
Urbanisation	Sweden, Bulgaria, Denmark, Hungary, Italy	Bulgaria, Hungary	Bulgaria
Urbanisation/ suburbanisation	Yugoslavia, Poland, Austria, Netherlands West Germany	Yugoslavia, Sweden, Poland, Italy, Austria, Denmark	Poland
Suburbanisation	Switzerland, Great Britain, Belgium	France, Switzerland, West Germany Netherlands, Great Britain	Austria, France, Italy, Denmark, Sweden
Suburbanisation/ de-urbanisation	–	Belgium	Great Britain, West Germany, Netherlands, Switzerland
De-urbanisation	–	–	Belgium

Source: Van den Berg (1982)

a process of re-concentration of higher level economic activity within the older city centres. Strengthened by urban revitalisation policies, a number of large cities in the 1990s, particularly in France, Germany and the Netherlands increased in population for the first time since their decline in the 1970s. This was often accompanied by the process of 'gentrification' (an inflow of relatively high income and mobile households) and related environmental enhancement (CEC, 1991b). By the late 1990s, half of the functional urban regions of the EU were growing, often with stable or increasing populations in their cores. In the United Kingdom, some central cities were declining less rapidly than in the 1980s, while urban regeneration policy helped both to increase household numbers and diversify the social-economic mix within certain parts of declining cities (Maclennan, 1997).

Re-urbanisation and gentrification

Within the urban core and inner suburban ring, gentrification is a problematic aspect of re-urbanisation. Suburbanisation in the past was associated with an out-migration of younger and economically active households, with older and less mobile households remaining behind in the inner cities. Although dis-investment and neglect produced a 'rent-gap' in many inner-city housing areas, making the rehabilitation of old housing profitable, it also meant that low-income residents would be replaced by more affluent tenants who were able to afford higher rents (Wegener, 1995). In the United Kingdom this process was often compounded by the provision of improvement grants, where, in London in the early 1970s, landlords and developers received 75 per cent of all grants awarded (Balchin, 1971).

At the top of the urban hierarchy, in London and Paris, and also in second rank cities such as Bruxelles, Frankfurt, Milano and München, re-urbanisation is highly inflationary, since large-scale real estate speculation and exorbitant increases in commercial rents make housing in the central areas of cities unaffordable to all except the very affluent. The same is also true in some Central European cities such as Budapest and Praha, where the inflow of multi-national companies have resulted in soaring office rents and the displacement of housing in core areas.

Wegener (1995) suggests that if these trends continue, the modern European metropolis might become divided into three different cities:

1   The 'international' city with airports, hotels, banks, offices, luxury flats and a prospering central shopping area.
2   The 'normal' city for native middle-income households – hidden behind the international city in low-density suburbs or in high rise housing areas on the urban periphery.
3   The 'marginalised' city for the poor, the elderly, the unemployed, and the migrant worker – in parts of the urban core but also in devalued peripheral areas.

Gentrification is likely to hasten division. If gentrification takes place within the international city, the poor will be pushed out into low-quality high-rise housing of the 1950s and 1960s within the marginalised city (Wegener, 1995).

Social exclusion

Although the tide of de-urbanisation in the EU has generally subsided and in many cases has turned 'before too many functions flowed from the city to the suburb in the wake of the population . . . [this] . . . was not always the case and the plight of the inner city is a very real one' (Burtenshaw *et al.*, 1991: 297). In so many inner cities, housing in a poor condition is often inhabited by low-income households, the unemployed or by recent immigrants, while there are also numerous examples of the inhabitants of a run-down inner suburb being transferred to devalued peripheral estates – thereby remaining within the marginalised sector of the urban economy. An urban crisis clearly exists and varies in its degree of intensity throughout the EU. In analysing the economic performance of 119 cities between 1974 and 1984, Cheshire and Hay (1989) showed that urban crises could occur in cities experiencing growth or decline, and differentiated between cities with 'significant' problems and those with 'serious' and 'severe' problems. Among thirty-eight cities with poor economic performance, twenty-four were in the process of decline, only eleven were likely to grow, and three were in a position whereby they might either decline or grow. In addition, their analysis showed that the economic performance of two cities was improving sufficiently to enable them to be classified as cities which would soon have significant problems rather than cities with serious problems, whilst the performance of two other cities was deteriorating to such an extent that they would soon face severe problems rather than serious problems (Table 11.4). It is of note that whereas the problems of decline are most evident in the older industrial cities of North-West Europe, the problems of growth are characteristic of Greece, southern Italy, central and northern Spain, and Portugal (Burtenshaw *et al.*, 1991).

By the mid-1990s, the European Commission recognised that social exclusion was a growing problem for most large cities in the EU whatever their economic attributes and wherever they are located. *Europe 2000 +* (CEC, 1994c) emphasised that although the scale of the problem varies markedly from one city to another, the manifestations were similar: concentrations of poor housing, increasing differences in health and life expectancy, growing disparities in educational and skill levels and access to decent jobs, high levels of long-term unemployment, widening income disparities and a rising crime rate. Whereas in the industrial cities of the North, the problem was concentrated in certain deprived areas in the inner city or the suburbs, in the South it is often more diffused reflecting the more chaotic nature of urban growth and the lack of planning controls.

Table 11.4 The typology of problem cities

Significant problems		Serious problems		Severe problems	
Cardiff	D	Birmingham	D	Belfast	D
Coventry	D	Hull	D	Charleroi	D
Derby	D	Lille	D	Glasgow	D
Genova	D	Messina	D	Liége	D
Le Havre	D	Newcastle	D	Liverpool	D
Leeds	D	Sheffield	D	Sunderland	D
Manchester	D	Le Havre	D	Valenciennes	D
Rotterdam	D	Bilbão	D/G	Teesside	D/G
Rouen	D				
St Etienne	D				
Teesside	D				
Barcelona	G	Dublin	G/D	Cagliari	G
Murcia	G			Cordoba	G
Palermo	G			Malaga	G
Valladolid	G			Napoli	G
Zaragoza	G			Sevilla	G
Granada	G				
←		Athinai	G		
←		Lisboa	G		
		Porto	G	→	
		Thessaloniki	G	→	

Source: Cheshire and Hay (1989)

Note: D = Decline; G = Growth

The European Commission subsequently reported that in the densely populated zones of the EU, 11.9 per cent of the workforce was unemployed in 1995, compared to 10.8 per cent in rural areas and 9 per cent in areas which were semi-urbanised though within easy access to highly urbanised centres. While some cities had comparatively low unemployment rates (for example, Frankfurt and Milano), others (such as Bruxelles, Birmingham, København, Köln, Napoli and Palermo) exceeded national or European averages by at least a fifth (CEC, 1997c). The Commission recognised that while the more highly qualified segment of the labour force was able to compete in an open economy, taking advantage of new economic opportunities in many cities, a more vulnerable group had emerged which lived in semi- or permanent exclusion – widening economic and social disparities. In cities, multiple deprivation is expressed in bad housing conditions, homelessness, social isolation, poor educational attainment, drug abuse and criminal behaviour, each of which render access to the labour market, at best, problematic. In consequence, in densely populated areas long-term unemployment amounted to 56 per cent of total unemployment in 1995 (CEC, 1997c).

Although social exclusion in Northern European cities has led for generations to the spatial segregation of social groups in neighbourhoods with poor facilities, this pattern is becoming increasingly evident in the cities of Southern Europe where unemployment rates in excess of 30 per cent are as common as in the north (CEC, 1997c). It is also clear that social exclusion often overlaps cultural and linguistic differences in many urban neighbour-hoods, while it is increasingly recognised that spatial segregation is not only a social problem in terms of poor housing, education and employment, 'but that socially deviant behaviour which results from segregation harms the economic attractiveness of the city' (CEC, 1997c).

Urban policy

Since their inception, the European Coal and Steel Community (ECSC) and the European Investment Bank (EIB) have been involved in regional devel-opment and have thus influenced the pace and nature of urban growth and change (Pinder, 1983; Burtenshaw *et al.*, 1991). However, until the 1990s, moves to develop a coherent EU approach to urban problems were con-strained by the priority given to agriculture. Although around 80 per cent of the EU's population lived in urban areas, up to two-thirds of the EU's total budget was allocated to farm price guarantees under the Common Agricultural Policy. But apart from regional aid being disproportionately low, the urban element of regional policy was regarded by the European Commission as a matter of insufficient concern to justify establishing a directorate general (DG) with exclusive competence for policy in this area. However, of the twenty-four DGs of the European Commission, six DGs (notably DGs V, VII, XI, XIII, XVI and XVII) have, from time to time, given an urban dimension to their specific policy sector (Williams, 1996). The DG whose competence is most relevant to the theme of this chapter is DG XVI, Regional Policy and Cohesion. Of the others which have respon-sibilities for policy in an urban context, DG XI, Environment and DGVII, Transport are the most important. Since DG XVI implements its urban policy through the medium of the Structural Fund (particularly the European Regional Development Fund and Community Initiatives) and the Cohesion Fund, this chapter will focus initially on fund-aided urban development, prior to a consideration of environmental and transport policy within an urban context.

Structural Fund and urban development

The European Regional Development Fund

From its creation in 1975 to the early 1990s, the ERDF was not specifically targeted at urban areas, although to an extent aid was inevitably channelled into towns and cities. However, although conferences of European Ministers of Planning at Bari in 1976 and Torremolinos in 1983 adopted specifically urban themes, there was an emphasis on regional issues throughout most of this period – a tendency reinforced by the accession of Greece, Portugal and Spain (Burtenshaw *et al.*, 1991). Since there was a lack of spatial congruence between the areas eligible for ERDF assistance and FURS with the worst economic, social and environmental problems, the ERDF clearly did not benefit urban areas in greatest need (Cheshire and Hay, 1989).

However, over the period 1994–99, it was recognised that the success of urban areas was central to the overall growth and development of Objective 1 regions, while the highly urban character of Objective 2 regions resulted in constituent urban development projects occupying a large share of ERDF assistance. Of the global ERDF budget of 60 billion ECU (accounting for 49.5 per cent of the resources of the Structure Funds, 1994–99), around 21 billion ECU was being spent on development within urban areas of at least 100,000 inhabitants (the EU definition of a city) (CEC, 1997c). By comparison, from the other Structural Funds (the European Social Fund, the European Agricultural Guidance and Guarantee Fund, and the Financial Instrument for Fisheries Guidance) very few or any financial resources were invested within the eligible urban areas.

In the Objective 1 regions, the urban component of ERDF allocations in the period 1994–99 ranged from as little as 19 per cent (in Northern Ireland) to as much as 100 per cent (in East-Berlin) (Table 11.5). The urban areas of Spain, Portugal and Greece received the largest amount of ERDF aid, equivalent to between 34 and 40 per cent of their total

Table 11.5 The urban component of ERDF allocations in Objective 1 regions, 1994–99

Country		*million ECUs*	*% Objective 1 funding*
Portugal		4,082	40
Spain		5,984	38
Greece		3,207	34
Germany		1,670	24
	(East-Berlin	530	100)
	(Rest of Eastern Germany	1,140	20–25)
Italy		1,000	30
Ireland		900	35
France		750	56
Belgium		515[1]	90
UK	(Merseyside	271[2]	90)
	(Northern Ireland	126	19)
Netherlands		80	37

Source: CEC (1996, 1997)

Notes
[1] Total Structural Fund allocation to Objective 1 regions in Belgium.
[2] Total ERDF support for Community Economic Development programmes in Merseyside and UK Objective 2 regions

Objective 1 allocations (CEC, 1997c). In Spain, investment took place in, for example, the extension of high quality digital telecommunication networks in all urban areas, the zonal development of urban industrial land, and the improvement of urban transport and the urban environment. In Portugal, the urban component was invested in industrial development in urban areas, the improvement of the urban infrastructure, and environmental and urban renovation including the improvement of living conditions in the barracas of Lisboa and Porto. In Greece, 70 per cent of the urban funding was targeted at Athinai, in which investment in the city's metro was financially the largest single undertaking in all EU assisted urban areas. The urban areas of Eastern Germany were also major recipients of Objective 1 aid. In East-Berlin, funds were invested in the redevelopment of numerous sites both in the heart of the city and in several of the city's industrial zones, while within the other *Länder* aid was targeted at a wide range of economic and environmental regeneration projects.

Middle-ranking recipients of Objective 1 urban aid comprised Italy, Ireland, France and Belgium. In Italy funds were targeted particularly at a range of economic and environmental regeneration projects in the cities of Napoli, Bari and Palermo, and more widely distributed among urban areas in Abruzzo, Calabria, Campania, Puglia, Sardegna and Sicilia; and in Ireland, apart from investment in the redevelopment of the Temple Bar area of Dublin and in the city's Light Rail Transport System, there was support in other urban areas for local enterprise measures, employment, physical renewal and economic regeneration. Objective 1 funding also assisted the economic regeneration of the urban centres of French Hainault – a region characterised by industrial decline; while in Belgium the cities of Charleroi, Mons and Louvier in the province of Hainault, similarly in decline, were the principal recipients of ERDF support (CEC, 1997c).

Northern Ireland, the Netherlands and Merseyside were the three smallest recipients of Objective 1 funding. In Northern Ireland support was given, for example, to a sewage treatment plant in Belfast and to the physical development of the Belfast Docks, while in the Flevoland region of the Netherlands nearly 40 per cent of Objective 1 aid was spent on urban development mainly in Almere and Lelystad (CEC, 1997). Within Merseyside (an Objective 1 region) and the Objective 2 areas of the UK, funding (amounting to 271 million ECU, 1994–99) was used to facilitate 'Community Economic Development (CED) in specific areas of exceptional deprivation. By focusing on pockets of social exclusion and long-term unemployment, CED was intended to involve local communities and businesses in the process of regeneration, and by reintroducing the most vulnerable groups in society into the regular economy, CED made a positive contribution to the broader aims of urban economic development (CEC, 1997).

Because of high population densities in the Objective 2 areas of Germany, the UK and Belgium (Table 11.6), cities were the principal recipients of

Table 11.6 National average population density of Objective 2 areas

Country [1]	Average population/km² in Objective 2 areas
Germany	911
UK	650
Belgium	484
Netherlands	472
Luxembourg	389
Spain	349
Italy	256
France	167
Denmark	88
Austria	73
Finland	46
Sweden	26
Total EU	245

Source: CEC (1997)

Note: [1] There are no Objective 2 areas in Ireland, Greece and Portugal

ERDF assistance in these localities – often accounting for at least 80 per cent of Objective 2 allocations. However, whereas in Germany and in the region around Liége in Belgium there was little variation in the extent of urbanisation (most urban areas received 80 to 100 per cent of allocations), in the UK there was a contrast between the many English Objective 2 areas (where, because of high densities, percentages of aid received by cities tended to be around 80 to 90 per cent) and those in eastern and western Scotland and South Wales (where lower densities resulted in the constituent urban areas receiving only 40 to 50 per cent of Objective 2 assistance). In Germany, Wallonia in Belgium, and the UK, industrial decline is very largely an urban phenomenon, and therefore Objective 2 funding in these countries was mainly targeted at urban redevelopment. Recipient projects tended to be either typical economic–regeneration ventures concerned with productive activity (for example, enterprise support measures), or of a facilitating nature (such as the provision of a modern transport infrastructure).

In the Netherlands, Luxembourg, Spain, Italy and France (with only moderate population densities in their eligible areas), Objective 2 funding had a variable impact on urban development. Whereas in the Netherlands, almost 100 per cent of Objective 2 aid benefited the highly urbanised area centred on Arnhem-Nijmegan, the financial impact on cities in Italy varied from 15 per cent in Lazio to 70 per cent in Liguria, while in France the impact ranged from 30 per cent (in Centre, Auvergne, and Languedoc-Roussillon) to 75 per cent (in Basse-Normandie, Champagne-Ardennes, Franche-Compté, Pays de Loire, and Poitou-Charentes) (CEC, 1997c). In Spain there was a similar variation of impact, with Barcelona and Bilbao having received the lion's share of aid.

Since the Objective 2 areas of Luxembourg, Austria, Finland and Sweden contained no cities with more than 100,000 inhabitants (the definition of an urban area, as used in the above analysis), it has to be assumed that Objective 2 funding in these countries was concentrated in smaller towns which served as growth poles for bigger geographical areas (CEC, 1997c). Clearly there was a different pattern of industrialisation and urbanisation than in other EU countries, and that, in the absence of large cities, industrial decline within the designated areas was an attribute of smaller towns and semi-rural areas.

Urban Pilot Projects

Jointly financed by the ERDF and local government, Urban Pilot Projects were designed to test new ideas in the application of urban policy. Focusing effort on those areas where they would be most effective, four categories of projects were supported during the first phase of operation commencing in 1989 (the last category being added in 1992).

1   Economic development in areas with social problems (for example inner-city and peripheral estates with high unemployment);
2   Environmental actions linked to economic goals;
3   Economic and commercial revitalisation of decayed historical centres;
4   Exploitation of the technological assets of cities.

A total of thirty-one projects were supported throughout the EU, with the greatest numbers being located in the decayed urban areas of cities in France, Germany and the United Kingdom (Table 11.7).

Although it is difficult to undertake an immediate evaluation of the effectiveness of Urban Development Projects, by the mid-1990s it was clear that ERDF funding enabled many projects to go ahead which otherwise might not have been undertaken (RECITE, 1995). It was evident that enterprises had been set up and employment generated, local partnerships had been established to meet vocational training needs, and redundant buildings and land were brought into beneficial use (Williams, 1996). It will be necessary to wait until the early years of the twenty-first century before the results of an appraisal of the second phase of projects (1995–99) are known.

Community Initiatives

The European Commission launched the Community Initiative URBAN in 1994 to improve and extend the co-ordination of the EU measures directed at urban problems and specifically to focus on the problems of spatial segregation in cities. The Commission was concerned that the unemployment and other socially vulnerable groups were becoming increasingly concentrated in specific neighbourhoods to worrying proportions in recent

Table 11.7 Urban Pilot Projects, 1989–94

Country	Project and host city			
	Economic development in areas with social problems	Environmental actions linked to economic goals	Revitalisation of historic centres	Exploitation of the technological assets of cities
France	Lyons	–	–	Bordeaux
	Marseille	–	–	Montpellier
	–	–	–	Toulouse
Germany	Bremen	Neuenkirchen	Berlin	–
	Dresden	–	–	–
United Kingdom	London	Belfast	–	–
	Paisley	Stoke-On-Trent	–	–
Belgium	Antwerp	–	–	–
	Bruxelles	–	–	–
	Liége	–	–	–
Spain	Bilbao	Madrid	–	Valladolid
Netherlands	Groningen	–	–	–
	Rotterdam	–	–	–
Denmark	Aarlborg	–	–	–
	København	–	–	–
Greece	–	Athinai	Thessaloniki	–
Ireland	–	–	Dublin	–
	–	–	Cork	–
Italy	–	–	Genova	–
Portugal	–	–	Lisboa	–
	–	–	Porto	–
Number of Projects	14	5	7	5

Source: RECITE (1994)

years (CEC, 1997c). With programmes implemented in approximately 115 cities (with populations of 100,000 or more), URBAN had a budget of around 850 million ECU (at 1996 prices) for the period 1994–99, of which two-thirds were channelled into Objective 1 areas.

URBAN aid specifically facilitated integrated development programmes in deprived areas of cities and was targeted at economic, social and environmental problems in areas with decaying infrastructure, poor housing and a lack of social amenities (CEC, 1994c). Based on the principle of subsidiarity, examples of typical measures undertaken by local partnerships included integrated programmes involved with the launching of new economic activities (for example, the provision of workshops and business centres), training schemes (such as the teaching of computer skills), the improvement of social, health and security provisions, and the improvement of infrastructure and the environment through the renovation of buildings and rehabilitation of public spaces (CEC, 1994c).

Following the introduction of URBAN, the European Commission launched INTEGRA as part of the Employment Community Initiative in 1995. Focusing on human resource development in deprived urban areas, it combined a local approach to neighbourhood regeneration with initiatives for job creation (CEC, 1997c). INTEGRA aimed to increase the awareness among the disadvantaged of an integrated approach which simultaneously aimed to resolve the multiple problems that faced people who were excluded from the labour market, such as health, housing, social protection, mobility, and access to justice and to public services (CEC, 1997c).

The Cohesive Fund

Applicable only to Greece, Ireland, Portugal and Spain, the Cohesive Fund helped to enhance the functioning of constituent conurbations as a whole, for example through investment in public transport schemes and environmental projects such as the reclamation of derelict urban land and the treatment of urban waste.

Financial resources from the Cohesion Fund were split 50/50 between transport and the environment. As far as transport was concerned, eligible projects (of which 71 per cent were road, and 21 per cent were rail) were to be connected to trans European networks (TENs) or were to feed TENs, and would thus have an important impact on urban areas. Clearly, the connection of the periphery of the EU to the core, and the consequential reduction in travel time, lowered both the production costs and distribution costs of enterprises in peripheral cities (CEC, 1997c). With regard to the environment, water supply and wastewater treatment absorbed 72 per cent of allotted Cohesion Fund resources, and waste treatment accounted for a further 4 per cent. Most of this expenditure was intended to enhance the functioning of urban areas (CEC, 1997c).

Urban transport policy

Within the EU, investment in new transport infrastructure has not kept pace with increasing volumes of traffic and has resulted in a worsening of congestion in and around the Community's densely populated core areas such as the Ile de France, Greater London, the Rhein-Ruhr area, München, Milano and Madrid.

Within these congested areas in particular, but also in smaller cities, the provision of efficient transport services is crucial for continuing urban development, since it is both 'a determining factor in the competitiveness of the urban economy and in the quality of life of city-dwellers' (CEC, 1997c: 9). With this in mind, increasingly governments within the EU (often advised by environmental pressure groups) are increasingly basing their transport policies on the notion of 'sustainable mobility' which aims to reconcile the demand for mobility with both the availability of resources and the impact of transport operations on the environment. In this respect, urban transport policy is being gradually designed to contribute to a reduction in the conflict between traffic and the environment.

To benefit all urban dwellers, public transport should not only provide an alternative means of travel to the private car, but also have an important contribution to make to social cohesion, particularly in areas where people without cars (such as younger or elderly people, and low income groups) need to have access to economic and social activities; while, for much the same reasons public policy should promote other alternatives such as cycling and walking. To further the formulation of policy along these lines, the Green Paper, *The Citizen's Network* (CEC, 1996) defined the main action areas at a Community level to encourage and promote an integrated, intermodal transport system which fully exploits the potential of public transport (CEC, 1997c). Clearly, to reduce congestion and environmental damage, appropriate pricing policies would need to be introduced to ensure a more rational allocation of resources between the various modes of urban transport. In this regard, public transport fares could be lowered by means of subsidisation, and the price of using the private car could be increased to reflect the social cost of road use. In some heavily congested urban areas, the use of the private car might be deterred by high parking charges, but increasingly city authorities are considering the introduction of charges for the use of urban roads.

On a wider geographical scale, TENs are instrumental in increasing access to cities, but it is important that investment in TENs is fully integrated with urban transport networks to ensure that both peripheral and urban areas both benefit from long distance links (CEC, 1997c).

Urban environmental policy

People in cities are increasingly concerned about the quality of their natural environment. In 1995, 70 to 80 per cent of European cities with more than 500,000 inhabitants failed to meet the World Health Organisation's air quality standards, while 70 million city dwellers in the EU were affected by 'winter smog' indicators often being twice the quality standard ceiling (CEC, 1997c). The rising concentration of cars in cities and the treatment of solid waste and wastewater also adversely affect the urban environment.

Although not laying the specific foundations for the development and application of remedial environmental policies, the ideas presented in the *Greenbook on the Urban Environment* (CEC, 1990b) were taken seriously by the many politicians and professionals concerned with urban planning (Williams, 1997). The document first examines the urban environment and the underlying causes of urban degradation, and second suggests a strategy and areas of action for the enhancement of the urban environment. Within an urban planning context, it focuses on the relationship between the development of the built environment and air pollution, pressures of suburbanisation, traffic calming, conservation of the cultural heritage, protection of historic buildings, green spaces and tourism.

Invoking the principles of policy co-ordination, subsidiarity and co-ordination, the areas for action involve policies which both concern the physical structure of the city (viz urban planning, urban transport, protection and enhancement of the historical heritage, and protection and enhancement of the natural areas within cities), and reduce the impact of urban activities on the environment (viz urban industry, the management of urban energy, urban waste management and urban water management).

Within the context of urban spatial policy, the *Green Book* proposed the adoption of a number of practices which in some member countries would have appeared, at best, controversial, for example: the development of mixed urban land-uses and higher densities (to reduce a dependency on the car), the re-use of abandoned, derelict and contaminated industrial land, and the revitalisation of social housing estates, and suggested that the European Commission (in co-operation with Member States) should ensure that environmental considerations are fully incorporated into urban planning strategies (Williams, 1997). While the latter has led to the introduction of strategic environmental assessment, some of the other proposals have been adopted in a number of Urban Pilot Projects.

Although the *Green Book* was arguably discredited since its proposals for action both encroached upon the responsibilities of regional and local government and concerned such matters over which the EU had no competence, it nevertheless led to the formation of the Urban Environment Expert Group by the Council of Ministers in 1991. The Group (consisting of representatives from all EU members) focused its attention on the issue of sustainable urban development – a Sustainable Cities project being established to review a range of initiatives emanating from international organisations such as UNCED's Local Agenda 21, OECD's Ecological City Project and the *Fifth Action Programme on the Environment* (CEC, 1992a; Williams, 1997).

The fifth action programme was particularly notable in developing the concept of car-free cities, an idea which soon led to the setting-up of the Car Free Cities Club in 1994 – membership being open to all cities in the EU. The main aim of the club was to 'establish the means whereby experience of controlling traffic in order to create a better environment could be more easily exchanged' (Williams, 1997: 209), and to promote a wide range of projects demonstrating different solutions to the conflict between urban traffic and the urban environment.

Continuing problems of urban growth

Within the context of the continuing integration of Europe, important economic and technical developments are taking place which will have a major impact on the pattern of urbanisation in many parts of the continent (see Wegener, 1995). The first of these developments is the continuing expansion of transport, which since the 1950s has been characterised by massive road construction, low fuel prices, a dramatic increase in the use of motor vehicles, the growth of intra-European and domestic air travel and the introduction of high speed train services. The resulting increase in accessibility has been a pre-requisite, particularly since the introduction of the Single Market in 1993, for both the unimpeded flow of goods and services throughout the EU and the break-up of spatial monopolies. Clearly, cities that are able to compete successfully will enjoy higher incomes and experience population growth, while those unable to compete will inevitably decline in terms of income and population – either relatively or absolutely.

The second development is the rapid growth in telecommunication, computerisation and automation with unquantifiable effects on the availability of products and the efficiency of factors of production in both the private and public sectors.

The third development is the inflow of immigrants from Central and Eastern Europe, reversing in some cases the process of de-urbanisation. Although initially, immigration might require the allocation of resources to meet housing, employment and welfare needs, in the long term the inflow of younger people might compensate for an increasingly ageing native population.

Because of the above developments, the urban system of Europe has become polarised between 'winners' and 'losers': between core and periphery, north and south, and east and west (Wegener, 1995). In the core, cities in south-east England, central and northern France, Belgium, the Netherland and West Germany have comparative economic advantages over cities in Scotland, Ireland, Greece and Portugal. In the north, urban population growth has virtually ceased, whereas in the south, where countries are still at an earlier stage of industrialisation and urbanisation, growth continues apace – although in some regions of Spain and in Italy economic development has been on a par with the north, birth rates have fallen, and urban growth is set to decline in the near future. In contrast, the east-west divide – wider than the north-south divide in terms of per capita income – might remain a more intractable division in the foreseeable future and become of increasing concern to the EU in its endeavours to expand eastward. However, while, integration might have a negative effect on much of the industrial economy of Poland, the Czech Republic and Hungary, since the Single Market could render it uncompetitive, cities such as Warszawa, Praha and Budapest could regain their pre-war status in the urban hierarchy of Europe.

Economic and political integration, together with advances in high speed transport, access to international airports, developments in telecommunications and computerisation, and large scale immigration will undoubtedly continue to have a major impact on urbanisation in Europe in the twenty-first century (Kunzmann and Wegener, 1991). With further advances in telecommunications, the growth of world markets and the continuation of major projects such as *les grand travaux* and docklands, the economies of the two global cities of Paris and London will become even stronger, while cities such as Bruxelles, Frankfurt/Main, Milano, Barcelona and Lyon will expand their transportation networks and create 'high profile' convention and cultural facilities (Wegener, 1995). Cities close to the interior border of the EU, for example Aachen and Strasbourg, might increasingly benefit from the Single European Market, while the opening of Central and Eastern Europe may enhance the position of cities which prior to the Second World War had access to East European markets (such as Hamburg, Frankfurt-an-der-Oder, København and Thessaloniki). As part of the process of de-urbanisation, important transport corridors will become the preferred location for market-oriented industry, distribution centres and households seeking lower house prices and a more desirable natural environment (Wegener, 1995).

Although the more affluent cities in Europe should be able to improve their transport and communication systems, poorer cities will suffer increasingly from ageing infrastructures, the lack of access to airports, and 'grey' area locations between high speed transport corridors (Wegener, 1995). Unemployment will continue to be a major cause for concern in the older industrial cities of the north, but will become increasingly severe in cities at the periphery, in ports that fail to modernise, and in cities formerly reliant upon high-tech defence related industries or military installations. Even cities dominated by the automobile industry, although prosperous at times of boom, could become severely disadvantaged during periods of recession. Perhaps of greatest concern are the comparatively low levels of prosperity in the cities of Central and Eastern Europe – where infrastructure is obsolete, the housing stock is below EU standards, environmental quality is poor and efficient systems of urban management have not, as yet, emerged (Wegener, 1995).

The adverse effects of polarisation in Europe, however, are not confined to cities in the poorer countries or regions. Rapidly rising land values in prosperous cities such as London, Paris, Madrid and München might continue to result in large areas of their inner suburbs becoming unaffordable to low income households, while the quality of their environment may be increasingly endangered by inadequate development control, growing traffic volumes, the lack of adequate ecological concern and (particularly in the growth areas of Southern Europe) insufficient public investment in sewage, waste disposal and energy systems (Wegener, 1995). Prosperity and poverty might become increasingly juxtaposed since, with the development of high-tech activity, more and more people will be left behind in the development of new skills and in their chances of securing employment. This, together with a tendency for governments to reduce expenditure on public housing and social security, might severely increase the degree of social exclusion in more and more cities in the EU. Within both poor and prosperous areas, the problem of social exclusion is compounded by the immigration of unskilled workers – for example in Paris, Frankfurt-am-Main and Amsterdam no less than in the 'gateway' cities of Frankfurt-an-der-Oder, Thessaloniki, Trieste and Marseilles where immigration from Eastern Europe or from Africa and the Middle East is likely to continue on a significant scale in the foreseeable future.

Future policy

The European Commission has belatedly recognised the role of cities as motors for regional, national and European economic progress, while at the same time has acknowledged that the depressed districts of medium-sized and larger cities have borne many of the social costs of past changes in terms of industrial adjustment and dereliction, inadequate housing, long-term unemployment, crime and social exclusion (CEC, 1997c: 13). Clearly, whereas there is a need to apply urban policies which ensure that cities remain at the forefront of an increasingly globalised and competitive economy, there is also an obligation to address the cumulative problems of urban deprivation. Unless this is undertaken, and solutions applied, economic progress will not only undermine the cohesiveness of urban areas but will also be unsustainable in terms of anti-social behaviour, rising crime rates and the absence of a social and political consensus for rapid change. There is thus an acknowledged need for an urban perspective in EU policies. In the view of the Commission (CEC, 1997c), to enhance the growth prospects of urban areas it is essential to explore the possibilities for:

1   expanding the TENs (particularly to ensure access to the networks from regional and local systems);
2   reinforcing intermodal freight and passenger transport;
3   developing telecommunication policies;
4   targeting research and technological development at the issues facing urban areas in the near future (for example integrated transport, energy, information networks, and sustainable construction technology and sustainable urban development);
5   consolidating the commercial functions of cities and their role; and
6   the introducing of clear targets for the improvement of the urban environment.

With regard to urban deprivation, the Commission considered that it was necessary to consider:

1   producing solutions to the problem of social and economic exclusion;
2   adopting a public health policy relevant to urban deprivation and poverty (and particularly to such issues such as poor housing and drug abuse);

3   dealing with matters relating to migration, the police and judicial co-operation, and crime; and
4   creating 'trust based' relationships between the various actors in the urban arena with the aim of promoting local empowerment, responsibility and initiative (CEC, 1997c: 15).

To ensure that the role of cities as growth points for regional development is enhanced, and that urban deprivation is alleviated, an integrated strategy of regional and urban policy is essential. This will necessitate local authorities participating closely in the preparation and implementation of regional development programmes and bringing in necessary expertise and knowledge to assist in the management of the local economy and labour market (CEC, 1997c). The Commission recognised that, even towards the end of the 1994–99 programming period, the following actions could have been undertaken:

1   focusing Structural Fund activities on pockets of high unemployment in the inner cities;
2   placing increased emphasis on public transport systems designed to increase the accessibility of peripheral regions, and to contribute to the resolution of the conflict between traffic and the urban environment;
3   prioritising the use of URBAN and INTEGRA programmes and reinforcing urban community development projects with the active participation of the local population.

It can be hoped that during the early years of the twenty-first century, the Commission will encourage the 'trans-national exchange of experience between cities with the objectives of collecting and compiling all relevant experience in urban regeneration and sustainable urban development, including the results of research in the socio-economic field' (CEC, 1997c: 17).

12 Conclusions

In the latter years of the twentieth century, regional policy and planning played a significant part in creating economic cohesion in the EU, but convergence on a Continental scale was, to an extent, offset by greater inequalities in prosperity between the rich and poorer areas of individual Member States. The EU manifestly exercised little influence over the effectiveness of the many planning systems employed at national, regional and local levels. Although most Member States were devolving planning responsibilities from the centre to the regions, devolution was a slow and long-term process and its attributes varied considerably from one country to another. However, at a European level, a representative Committee of Regions was formed in an attempt to voice matters of regional concern and influence decision-making in the European Parliament at Strasbourg, while, within a framework of super regions, Member States gradually embarked on strategic planning across national boundaries. It should not be forgotten, however, that 80 per cent of the population of the EU live in a polarised hierarchy of towns and cities, and that urban areas often need to be the focus of regional planning and (within the context of regional plans) the principal recipients of aid.

In the last two years of the century Member States addressed three substantial matters of concern, each of which, will have a major impact on regional policy and planning after the year 2000. First the reform of Structural Funds will inevitably mean that there will be many losers as well as winners throughout the regions of the EU. Second, the formation of the European Monetary Union and the adoption of a single currency will exacerbate regional inequalities in prosperity (unless there are adequate compensatory mechanisms), and finally the enlargement of the EU will increase the need to divert funds from the poorer regions of Western and Southern Europe to the even poorer areas of East Central Europe.

Convergence and divergence

As a result of market forces and the economic policies employed by Member States and the EU, the north-south income divide in Europe is narrowing, but the gap between the rich and poor regions of individual countries is widening, according to the *Report on Economic Cohesion* (CEC, 1996a). The report revealed that over the period 1983–95, disparities in the standard of living of Member States narrowed significantly. This was largely attributable to the GDP per capita of the cohesion countries – Ireland, Spain, Portugal and Greece – catching up with the EU average as a result of comparatively rapid economic growth. Ireland made the most spectacular advance, increasing its GDP per capita from 63.6 per cent of the EU average in 1983 to 89.9 per cent in 1995, whereas Spain moved up from 70.5 per cent to 76.2 per cent and Portugal climbed from 55.1 per cent to 68.4 per cent, while Greece raised its income more modestly from 61.9 per cent to 64.3 per cent (Table 12.1). In contrast, Sweden and Finland lost ground compared to the rest.

Across the EU, experience with regard to employment was mixed. Job creation in the UK, West Germany, the Netherlands, Belgium and Portugal, although variable, was sufficient to reduce unemployment, whereas in the country with the highest rate of economic growth, Ireland, employment grew by only 0.2 per cent from 1983 to 1993. In Finland and Sweden, because of deep recession, there was an absolute decline in employment over the same period, while rates of unemployment not only soared dramatically in these countries but also increased substantially in the cohesion countries, Spain and Greece (CEC, 1996a).

In considering income disparities across the regions of the EU, there are indications that these remained largely unchanged over the period 1983–93. In the twenty-five most prosperous regions, GDP per capita rose marginally from 140 per cent to 142 per cent of the EU average, whereas in the poorest regions it increased from 53 per cent to 55 per cent, and in Objective 1 regions it increased from 65 per cent to 67 per cent (CEC, 1996).

However, when examining regional income disparities in Member States, it is evident that, with the exception of the Netherlands, disparities widened markedly. Hamburg retained its position as the Union's wealthiest region with 189 per cent of average EU GDP per capita, Bruxelles (183 per cent) was second, changing places with the Ile de France (163 per cent), and Greater London (144 per cent) fell from seventh to ninth place. The poorest

Table 12.1 Gross domestic product per capita – cohesion countries, 1983–95

Country	*(EU=100)*			
	1983	*1988*	*1993*	*1995*
Ireland	63.6	65.0	80.2	89.9
Spain	70.5	72.4	77.8	76.2
Portugal	55.1	56.5	68.2	68.4
Greece	61.9	59.6	64.5	64.3

Source: CEC, 1996

regions included Sachsen with 53 per cent of the average, Galicia (60 per cent), Andulucia (58 per cent) and Cantabria (61 per cent). Similarly, in all Member States, with the exception of the UK, regional differences in levels of unemployment also increased – in France and (West) Germany going hand-in-hand with a more unequal distribution of personal income. Across the EU as a whole, the incidence of unemployment became much more uneven. The twenty-five regions with the highest rates of unemployment increased their average level of unemployment from 17.2 per cent to 22.4 per cent over the period 1983–93, while the twenty-five regions with the lowest rates were able to reduce their average rate from 4.8 per cent to 4.6 per cent (CEC, 1996a).

While the primary responsibility for improving economic and social cohesion falls on the Member States, EU structural policies have nevertheless played an increasingly important role in income equalisation and job creation in recent years. Whereas, in the period 1989–93, an overall income equalisation (in terms of GDP per capita) of 3 per cent was achieved as a result of transfers of 0.3 per cent of EU GDP, it has been estimated that in the period 1994–99 an equalisation of 5 per cent resulted from transfers of 0.45 per cent (CEC, 1996a). However, the primary purpose of Community transfers is not to redistribute money per se, but, through investment, to strengthen the economic bases in recipient regions and, in consequence, to narrow the gaps between poorer and richer Member States. In the four cohesion countries for example, Structural Fund assistance in the period 1989–93 increased growth from 1.7 to 2.2 per cent per annum, and it has been estimated that in the period 1994–99 a further 0.5 per cent rate of growth will have been achieved, while in the poorer regions of many of the richer Member States economic opportunities have been exploited with the assistance of Union structural policy – for example, 530,000 jobs were created or maintained in Objective 2 regions in the period 1989–93, and an estimated 500,000 jobs were created or maintained in Objective 5b regions in 1989–99. In addition, Community Initiatives have targeted European problems, identified new opportunities for development and improved inter-regional and cross-border relations.

Devolution and planning

Although it is apparent that in several Member States there is an increasing degree of devolution of central government planning power to the regions (see Chapters 3–8), at a local level there remains a considerable diversity in systems of planning – varying at the one extreme from a land-use focus which exists within the UK, to a competence in regional, environmental, transport, energy and communications policies in, for example, Germany. There has been, moreover, little attempt to harmonise systems throughout the Union. Only in a few cases, such as the requirement to undertake environmental impact assessments for certain development projects, have EU regulations impacted upon planning practice (Healy, 1997: 13).

However, with the incorporation of town and country planning into the Maastricht Treaty under the 'environmental' title, the European Commission is now empowered to act within this area if it deems it necessary to do so. Initially, the Commission has focused its attention on Europe's environmental agenda – issuing directives on air quality and on landfill, but the planning systems of individual Member States will soon come under the influence of European legislation if proposed directives such as the *Assessment of the Effects of Certain Plans and Programmes on the Environment*, are adopted (Healy, 1997: 13). Under this proposal, all statutory land-use plans within the EU will be subject to a strategic environmental assessment before adoption. The *European Spatial Development Perspective* (CEC, 1998a) will also affect planners across the EU. Promoted as a means of supporting more balanced territorial development across Europe, the ESDP brings together a number of different policies and identifies where investment is needed. In particular, it aims to bring about 'a more balanced system of towns and cities; parity of access to infrastructure and a prudent management of the heritage' (see Nadin, 1998: 62). In contrast to the UK system of land-use planning (involving the preparation of structure plans, local plans, unitary development plans and minerals local plans), the ESDP is based on 'integrated systems adopted in other Member States, incorporating investment commitments and bids for programme and project funding' (Morphet, 1997: 122–3). If the Perspective influences EU policy, some Members States (including the UK) will need to adopt more integrated systems of planning than hitherto if they are to make successful bids for Structural Funding. Integrated planning, however, will almost certainly need to be on a regional scale. This is a matter of concern particularly in the UK since it is in this country that the *Compendium of European Spatial Planning* (CEC, 1998a) identified regional planning as about the weakest in the EU. However, there are signs that with the devolution of planning powers to Scotland, Wales, Northern Ireland and Greater London, and with the development of Regional Development Agencies, integrated regional planning in the UK might be on the ascendancy.

Regional representation

To enable regional and lower-tier authorities to participate in an advisory capacity in the EU's decision-making processes, a Committee of Regions (CoR) was established by the Council of Ministers as a direct outcome of the Treaty of Maastricht. Nominated by national governments, the 222 members and 222 alternates of the CoR are representatives of regional and local government (but in the case of the UK, where there was an absence of an elected-tier of regional government, all twenty-four members and their alternates were appointed from among local government councillors). Care was taken to ensure that a geographically-balanced delegation of members was selected, including representatives of the major cities (Table 12.2), and, once appointed to the CoR, membership continued for five years.

The CoR is consulted by both the European Commission and European Council on matters such as regional policy, the Cohesion Fund or the Trans-European Network, particularly if, through the process of subsidiarity, regional and lower-tier authorities are responsible for implementing the relevant EU directives and regulations after they had been adopted. Based in Bruxelles, its work is undertaken by thirteen commissions and

Table 12.2 The Committee of Regions: membership, 1993–94

Country	Number of seats	Representation	
Belgium	12	Vlaams Gewest	5 seats
		Région Wallonne	2 seats
		Bruxelles	2 seats
		Wallonne community	2 seats
		German-speaking community	1 seat
Denmark	9	Amtere	4 seats
		Municipalities	4 seats
		København	1 seat
France	24	Balanced choice between 22 régions, départements and communes, allowing 8 for each level	
Germany	24	*Länder* choice, but 3 seats reserved for municipalities	
Greece	12	All nominated by government (8 mayors + 4 state officials)	
Ireland	9	All nominated by government	
Italy	24	Regioni	12 seats
		Provincie	5 seats
		Communes	7 seats
Luxembourg	6	Government choice from local authority nominees	
Netherlands	12	Municipalities (mayors)	6 seats
		Provincies	6 seats
Portugal	12	All nominated by government for 10 mainland seats and 2 for Madeira and Açores	
Spain	21	Communidadas autonomas	17 seats
		Local authorities	4 seats
UK	24	All nominated by government from elected local authority representatives:	
		England	14 seats
		Scotland	5 seats
		Wales	3 seats
		Northern Ireland	2 seats

Source: Middlemas (1995)

sub-commissions, with spatial planning being the concern of Commissions 1–5 and Commission 8 (Table 12.3).

The CoR, in addition and on its own initiative, has developed opinions on such matters as environmental protection and urban policy, and since spatial planning (involving a mix of town and country planning, infrastructure planning and economic development) is not a specific competence of the Treaty, the CoR supports the view that Member States should integrate the spatial aspects of Community policy and ensure that they are compatible with national and regional spatial planning strategies.

It can be questioned, however, whether the CoR adequately reflects regional interests and concerns. At the Maastricht IGC, the Commission had been in favour of universal election, but Member States (responsible for distributing 90 per cent of the regional budget) ensured that they alone should determine how the regions should be represented at Bruxelles (Table 14.3). The federal government of Germany, for example, was willing to permit its *Länder* a unimpeded representative voice on the CoR, whereas the Spanish government attempted to counterbalance the aspirations of the *communidades autonomas*, while the French government was often inclined to give *préfects* rather than its CoR members the principal responsibility for formalising regional relations with the Commission (Middlemas, 1995). Nevertheless, giving a voice to the regional dimension, the CoR rapidly became involved in the decision-making process of the EU during its early years of existence, and with extended powers its political authority could ensure that, in matters of regional policy and planning, its opinions become increasingly influential.

Table 12.3 Commissions and sub-commissions of the Committee of Regions

Commission		Sub-commissions
1	Regional development, economic development and local and regional finance	Local and regional finance
2	Spatial planning, agriculture, hunting, fisheries, forestry, marine, environment and upland areas	Tourism and rural areas
3	Transport and communications networks	Telecommunications
4	Urban policies	
5	Land-use planning, environment and energy	
6	Education and training	
7	Citizen's Europe, research, culture, youth and consumers	Youth and sport
8	Economic and social cohesion, social policy and public health	Institutional affairs [1]

Source: Williams (1996)

Note: [1] Special commission

Super regions

Clearly, greater economic and political integration in Europe will strengthen relations between regions in different Member States, while it could be increasingly argued that 'national boundaries are artificial and that the Europe of the future will be a federal entity based upon regions' (Gripaios and Mangles, 1993: 745). In the document *Europe 2000* (CEC, 1991b), attention was focused on the need for a bottom-up approach to the development of relations between regions and the encouragement of new ways of thinking which transcended national frontiers. The Commission therefore launched a series of research studies on eight regional groupings within the EU based on the criteria of geographical proximity and the degree of interrelationship between areas within them (see Figure 12.1). The territorial groups distinguished were:

- The Atlantic Arc: stretching from the north of Scotland to the south of Portugal;
- The Central Capitals: containing six of the capital cities of the EU;
- The Alpine Arc: comprising central-eastern France, southern Germany, the north of Italy (and Switzerland and Austria which were candidates for membership when the initial list as being drawn up);
- The western Mediterranean regions: in Spain, France and Italy;
- The central areas of the Mediterranean: Italian Mezzogiorno and Greece;
- The North Sea coastal regions: in the UK, the Netherlands, Germany and Denmark;
- The Continental Diagonal: inland parts of south-west France, and north and central Spain;
- The new German *Länder*.

In addition, a further super region was defined by a study entitled *Vision and Strategies around the Baltic Sea 2010: Towards a Framework for Spatial development in the Baltic Sea Region* (Group of Focal Points, 1994). The region embraced all countries with a Baltic sea coast, plus Norway and Belarus, and thus included both old and new EU Member States, EFTA and former Soviet and Visegrad countries and spanned wide disparities of wealth, living standards and environmental quality (Williams, 1996).

The European super regions, so defined, displayed considerable demographic and economic disparity. Omitting the new German *Länder* and the Baltic Sea Region because of lack of reliable data, Table 12.4 shows that the biggest super regions in population terms are the North Sea coastal regions and the Central Capitals each with 21.4 per cent of the EU population, while the smallest area is the Continental Diagonal with only 5.1 per cent. Annual population change ranged from 0.68 per cent in the western Mediterranean to as little as 0.10 per cent in the North Sea coastal regions, while population densities were greatest in the North

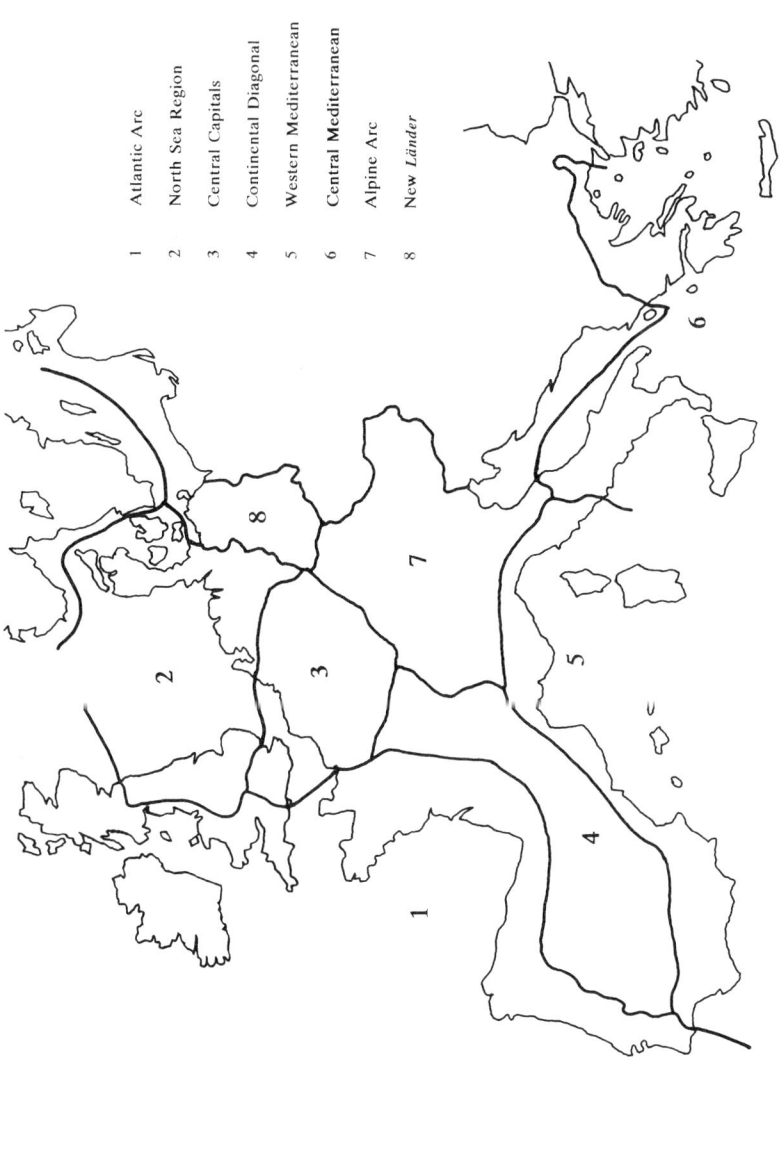

1 Atlantic Arc
2 North Sea Region
3 Central Capitals
4 Continental Diagonal
5 Western Mediterranean
6 Central Mediterranean
7 Alpine Arc
8 New Länder

Figure 12.1 Proposed super regions, EU

Table 12.4 The European super regions – as defined in 1991

	EC	Atlantic Arc	Central Capitals	Alpine Arc	Western Mediterranean Regions	Central Areas of the Mediterranean	North Sea Coastal Regions	Continental Diagonal
Population 1988 % of EC	14.9	21.4	16.5	11.2	9.8	21.4	5.1	100
Population change: 1978–88 average[1]	0.55	0.22	0.14	0.68	0.41	0.10	0.42	0.31
Population density 1988[1]	129	368	171	447	134	494	96	285
Population aged 15–64% 1987[1]	64	68	69	66	65	68	65	67
Participation rate 1990 average[1]	43	44	48	39	41	47	39	44
Unemployment rate 1990 averaged[1]	10.2	7.9	4.1	13.5	10.5	6.7	11.8	8.6
% employed in agriculture 1987[1]	16.0	5.0	7.5	9.0	29.3	4.1	16.2	11.4
% employed in industry 1987[1]	30	32	40	27	23	35	31	32
GDP per capita[1]	92	74	102	114	89	60	108	79
GDP per employee[1]	81	109	101	102	65	106	83	96

Source: CEC (1991b).

Note: 1. Unweighted averages of constituent NUTS-2 or, in the case of the UK, NUTS-1 regions.

Sea coastal regions and lowest in the Continental Diagonal. In terms of population aged 15–64, there was comparatively little variation, although the participation rate is highest in the Alpine Arc and North Sea coastal regions (reflecting employment opportunities and possibly social attitudes), while the lowest participation rates are in the Continental Diagonal and western Mediterranean regions (Gripaios and Mangles, 1993).

Economic disparities are no less marked. Unemployment rates ranged from as little as 4.1 per cent in the Alpine Arc to 13.5 per cent in the western Mediterranean. The proportion of the population employed in agriculture ranged from only 4.1 per cent in the North Sea coastal regions to as much as 29.3 per cent in the central Mediterranean, while the proportion employed in industry ranged from as little as 23 per cent in the central Mediterranean to 40 per cent in the Alpine Arc. GDP per capita also varied substantially – being highest in the Alpine Arc and lowest in the central Mediterranean, but if GDP per employee is considered the Central Capitals and North Sea coastal regions are the most prosperous areas.

Though the eight super regions of the EU are defined according to the criteria of 'geographical proximity' and their potential for 'developing mutual relationships', they are, however, far from being coherent entities in economic terms. In respect of the majority of selected economic variables, Gripaios and Mangle (1993) show that the average variation by sub-region is higher than by country – the only exception, and then only marginally, being the variables of employment in industry and GDP per capita (Table 12.5). By also analysing variations over the eight economic variables by super region and country, Gripaios and Mangle reveal that total variation within the European super regions is greater than the respective figure for countries. The above analyses suggest that the super regions as defined by the Commission are far from economically coherent.

If research suggests that there are contiguous geographic and economic regions, it would seem reasonable to develop specific policies for their

Table 12.5 Average variation of key indicators within super regions and countries

	Super regions	*Countries*
Population change, 1978–88	0.125	0.098
Population aged 15–64, 1987	3.597	1.662
Participation rate, 1988	13.951	7.670
Unemployment rate, 1990	18.428	15.358
% employment 1987 in:		
agriculture	55.639	47.217
industry	43.502	45.762
GDP 1986–8 (average)		
per capita	380.530	394.001
per employee	413.203	297.005

Source: Gripaios and Mangle (1993)

development, irrespective of national boundaries. Alternatively, policies could be developed to benefit non-contiguous regions defined on the basis of common socio-economic attributes, for example city states or high-tech locations – as already recognised by Community Initiatives such as URBAN (which is concerned with urban problems) and RECHAR (which facilitates the conversion of coalmining areas).

Despite the above criticisms and proposals, a number of Member States began work on the preparation of super regional plans in expectation that they would provide the basis for EU funding after 1999. In the United Kingdom, component plans for the newly-defined Atlantic Area, North Sea Region and the overlapping North West Metropolitan Area (incorporating the whole of the UK and Republic of Ireland) were at an early stage of preparation in 1998, but in Denmark, Sweden, Finland and Germany, component plans for the Baltic Sea super region were well advanced and identified areas which required investment, areas which were lagging or had growth problems, and areas of high environmental quality (Morphet, 1997).

Cross-border planning

It is within the broad framework of super regions that cross-border co-operation is becoming a major element in achieving cohesion within the EU. It is an approach which integrates 'bottom up' initiatives at the local and inter-regional level with 'top-down' policy at inter-state and EU level (CEC, 1994a). Motivated by the SEM and the Maastricht Treaty, and funded by INTERREG and INTERREG II, cross-border initiatives aim to achieve integration and harmonisation along the internal (and in some cases external) borders of the EU. In East Central Europe, similar developments are emerging – funded by INTERREG II and PHARE.

Border regions are often peripheral to the economic heartlands of their respective countries and suffer from a poor infrastructure, particularly if they are on the outer edge of the EU. Unemployment is often notably higher and GDP per capita normally substantially lower than the national average. Cross-border co-operation is clearly impeded by a wide range of socio-economic problems, notably monetary disparities between different countries, a lack of harmonised legislation, structural differences and unequal development, insufficient employment stability and a poorly co-ordinated economic structure. There are also administrative and cultural obstacles to co-operation, including different institutional structures, unequal local and regional government competences, an absence of trans-border transport and communication systems, and language differences (Vanhove and Klassen, 1987).

In the view of the CEC (1991b), cross-border co-operation is primarily dependent upon the completion of SEM, investment in infrastructure, and administrative and legal changes. The Commission had clearly recognised that it needed to achieve integration and harmonisation along its internal borders, support the development of the peripheral regions of the EU and promote good cross-border relations, and thus introduced the INTERREG programme in 1990 in an attempt to realise these aims. From 1990 to 1993, by means of thirty-one operational programmes and a budget of about ECU 1 billion, INTERREG helped to fund developments in transport and communications, trade and tourism, the environment and the rural economy. INTERREG II – with a budget of ECU 2.4 billion – focused on internal and external cross-border co-operation, particularly in respect of energy networks.

In the cross-border regions of the ECE countries, and particularly in those regions on both sides of the eastern boundary of the EU, socio-economic, administrative and cultural problems are often more severe than in the border regions exclusively within the Member States of the EU. Labour, capital mobility and the free flow of goods and services are generally restricted, transport and communication systems are less integrated, and cross-border problems of environment degregation need to be resolved. It is these problems that INTERREG II (in part) and PHARE were designed to address.

Before enlargement in 1995, there were 100,000km of frontier land of which 60 per cent was intra-EU and 40 per cent external (CEC, 1994a). Of the many terrestrial cross-border regions, the Rhein-Meuse-Nord region across the frontiers of the Netherlands and Germany is perhaps the most notable, while among the trans-maritime initiatives, the Transmanche region, the Welsh-Irish cross-border region, and the Öresund region are examples of very different forms of co-operation. On the external border of the EU, the largest initiative is along the German-Polish frontier and within the ECE countries numerous INTERREG II and PHARE programmes are in operation. These in turn are examined below.

The Rhein-Meuse-Nord cross-border region

One of five Dutch-German cross-border initiatives, the Rhein-Meuse-Nord region (designated in 1984) embraces part of the borderland of the Nordrhein-Westfalen region of Germany and the Oost-Nederland region and lies within the Central Capitals super region. Table 12.6 shows Nordrhein-Westfalen has a larger area, larger population and a higher population density than Oost-Nederland, while in the 1990s the latter region had a higher proportion of its workforce employed in agriculture and services than its more urban neighbour, and although Oost-Nederland had a lower level of unemployment than Nordrhein-Westfalen it also had a smaller GDP per capita. In the ten years after 1984, there was little change in the comparative economic attributes of both regions, although there was a common drift away from agriculture and industry to service employment, while levels of unemployment in both regions increased dramatically.

With a population of 1.7 million people and an area of 3500 km^2, the Rhein-Meuse-Nord region is strategically situated between the ports of

Table 12.6 Principal demographic and economic attributes of Nordrhein-Westfalen and Oost-Nederland

	1994		1995			Unemployment	GDP
	Area (km2)	Population (000s)	Persons/ km2	Employment, % in				
				Agriculture	Industry	Services	%	per capita (EU=100)
Nordrhein-Westfalen	34,072	17,788	522	1.9	36.0	62.1	8.2	112
Oost-Nederland	10,495	3,163	301	5.0	25.3	66.2	7.1	93

Source: Office for National Statistics (1997)

Antwerpen and Rotterdam and the German agglomerations of Rhein-Ruhr and Rhein-Main. It is envisaged that through the co-operation of German and Dutch local and regional authorities and Chambers of Commerce on either side of the border, infrastructural, socio-economic and cultural development will be both stimulated and co-ordinated, alternative employment will be generated, networking developed and a unified economic area established. However, the extent to which these aims will be achieved depends upon whether certain obstacles to cross-border planning can be removed. There are, for example, two different planning systems (see Chapters 6 and 8), two different languages, two different labour markets, two different systems of taxation and social security, and an industrial information and co-operation gap (Mainstone, 1997). With the Anholt Treaty of 1993 conferring legal status on the region, the constituent authorities are now in a stronger position to collaborate in cross-border planning and to seek – as far as possible – a convergence of labour markets, reduced differences in taxation and social security systems, and greater industrial co-operation.

The Transmanche cross-border region

At the end of the twentieth century, the economy of the Transmanche region of Kent in England and the much larger *région* of Nord-Pas-de-Calais of France was markedly diversified – notwithstanding its location within the Central Cities super region. Kent was economically stronger than its French counterpart – having proportionately a larger service sector, a lower level of unemployment and a higher GDP per capita (Table 12.7). Kent, however, had substantial economic disparities locally. The relative prosperity of the county's industrial north-west in close proximity to London contrasted with depressed districts of eastern Kent. Nord-Pas-de-Calais, on the other hand, suffered from a more spatially uniform decline in heavy industry – losing 419,000 jobs between 1960 and 1990 – and was blighted by 10,000 ha of industrial dereliction and high levels of pollution (CEC, 1992).

Co-operation between Kent and Pas-de-Calais became a reality following the designation of the Transmanche cross-border region by the Co-operation Agreement of 1987. It was driven by the Channel Tunnel project and the decline of the tourist industry, and by the perceived need to co-operate in respect of the development of commerce, tourism and education. INTER-REG I funding became available in 1992, and was targeted at appropriate projects in Ashford, Canterbury, Dover, Shepway and Thanet in eastern Kent, and at projects in Calais, Dunkerque and Boulogne in France.

Building on the success of cross-border planning within the Transmanche region, INTERREG II funding for the period 1995–99 placed increased importance on strategic planning and economic strategy, and thus a larger Euroregion was established comprising Kent, Nord-Pas-de-Calais, Région

Table 12.7 Principal demographic and economic attributes of the Transmanche cross-border region

| | Area (km2) | Population (000s) | Persons/ km2 | 1990 Employment, % in | | | Unemployment | GDP |
				Agriculture	Industry	Services	%	per capita* (EU=100)
Kent	3.7	1,526	409	3	27	70	3.9	98
Nord-Pas-de-Calais	12.4	3,967	320	4	33	63	11.8	87

*1989

Source: European Commission (1992b)

Wallonne, Bruxelles and Vlaams Gewest. Spatial planning within the larger region is of particular importance, since while its population (15.5 million) amounts to 4.5 per cent of the EU total, the region embraces only 2 per cent of its area. Joint policies will focus on demographic change, economic opportunities, the integration of transport, an adherence to an agreed environmental charter, and on spatial planning issues such as urbanisation, historic town centres and urban sprawl (Williams, 1996).

The Welsh and Irish cross-border region

Located in the Atlantic Arc super region, the Welsh and Irish cross-border region comprises the western Welsh counties of Pembrokeshire, Carmarthen- shire, Ceredignon, Gwynedd, Conwy and Anglesey, and the south-east and mid-east of Ireland. Aggregate statistics for the whole of Wales and Ireland provide a broad indication of the demographic and economic attributes of the two constituent parts of the region. Table 12.8 shows that in the mid-1990s Ireland had a lower population density, larger proportion of its working population employed in agriculture and a higher level of unemployment than Wales, but nevertheless enjoyed a higher GDP per capita – although both areas, by EU standards, had a below average standard of living – the *raison d'être* for cross-border co-operation and INTERREG II funding.

Cross-border co-operation is deemed necessary for the innovation and development of markets, and the development of commercial ports and tourism on both sides of the Irish Sea. Under INTERREG II funding, Wales received ECU 12 million and Ireland ECU 62 million over the period 1995–99 – Wales and Ireland receiving grants of up to 50 and 75 per cent respectively. Future co-operation will probably involve joint projects concerned with monitoring the coastal and marine environment, the further development of export services for small and medium-sized enterprises, the development of tourism focusing on the Celtic and cultural links between the two nations, and cross-border training in information technology (Mainstone, 1997).

Although there are indications that the economies of Wales and Ireland have been improving in recent years, development is constrained by loca- tion. As part of the Atlantic Arc, the cross-border region suffers from an unbalanced urban system (there is only one large city – Dublin), it is isolated from major transport networks (and remote from the Channel Tunnel), productivity is low in agriculture, manufacturing industry is in decline and tourism is under-developed, and even when the Trans-European Network is complete its benefits will be spatially very uneven (CEC, 1994a). The Commission has suggested that the super region as a whole requires improvements in innovation, transport, communications, industry and tourism and in particular the development of technology networks. There is also a need to promote the development of medium-sized towns to reduce spatial disparities in economic opportunity.

Table 12.8 Principal demographic and economic attributes of Wales and Ireland

	1994			1995				Unemployment	GDP
	Area (km2)	Population (000s)	Persons/ km2	Employment, % in				%	per capita (EU=100)
				Agriculture	Industry	Services			
Wales	20,766	2,913	140.3	3.6	30.0	65.6		8.7	81
Ireland	68,895	3,587	52.1	12.0	27.7	60.0		14.3	88

Source: Office for National Statistics (1997)

Öresund

Separating the Kattegat and North Sea from the Baltic, Öresund has imposed locational disadvantages on the Swedish economy in terms of access to the principal markets of western Europe. Although Sweden is larger than Denmark in respect of area and population, has more natural resources, and a broadly similar employment structure than its neighbour, by the mid-1990s it suffered from higher unemployment and a lower GDP per capita (Table 12.9). Thus, in 1991, when Denmark and Sweden signed a treaty to construct a 15.5 km length four-lane motorway and double track railway from København to Malmö across the Öresund (by means of a bridge and tunnel), it was anticipated that not only would the locational disadvantages of Sweden be reduced, but that the Öresund region would become the principal urbanised area in northern Europe (Newman and Thornley, 1996, Williams, 1996). This latter concept was formalised in *Denmark Towards the Year 2018* (Ministry of the Environment, 1992) and further developed in *The Öresund Region – a Europole* (Ministry of the Environment, 1993).

Coinciding with the commuting areas around København and Malmö, the Öresund region is promoted by the regional and local authorities in both countries through the medium of the Öresund Committee set up in 1993. Apart from being the largest densely populated area in Scandinavia, the Öresund region boasts a high quality environment, a rich culture and strong knowledge base – including the resources of the universities and medical schools of København, Roskilde, Malmö and Lund (Newman and Thornley, 1996). With construction work starting in 1993, it was forecast that the project would be completed in the year 2000.

However, by 1998, the newly completed bridge and tunnel under the Storebaelt – linking Jylland with Funen and Zealand – was fully open for road and rail transport enhancing the locational advantages of København. Together with the Öresund bridge and tunnel, the Storebaelt project will, for the first time, link Sweden directly to the mainland of western Europe – transforming Malmö 'from a Swedish backwater into the country's European gateway' (Newman and Thornley, 1996: 236). The Öresund and Storebaelt projects will thus not only facilitate the development of a cross-border region centred on København and Malmö, but will link the North Sea Coastal super region with Sweden, Norway and Finland.

Cross-border co-operation in East Central Europe

During the period of Communist government in East Central Europe, the border areas along the western frontier were situated in some of the most undeveloped and backward regions on the Continent. Because of military and political restrictions, border areas suffered from economic discrimination and in many places the traditional ties between regions and cities were broken. The border areas were characterised by the presence of military

Table 12.9 Principal demographic and economic attributes of Denmark and Sweden

| | 1994 | | | 1995 | | | | Unemployment | GDP |
| | Area (km2) | Population (000s) | Persons/ km2 | Employment, % in | | | | % | per capita (EU=100) |
				Agriculture	Industry	Services			
Denmark	43,080	5,205	120.8	4.4	27.0	68.4		7.1	114
Sweden	410,934	8,780	21.4	3.3	25.8	70.9		9.1	98

Source: Office for National Statistics (1997)

defence zones, an undeveloped economy and a small number of border crossings.

Political change in 1989, the removal of the Iron Curtain, and the political and economic re-orientation towards western Europe created a new contextual framework for the development of border areas. The new circumstances were favourable for East-West co-operation, which could also take the form of cross-border co-operation. Several initiatives for the development of frontier zones and the promotion of cross-border co-operation are currently being undertaken. The western frontiers, however, are not only boundaries between two countries, but also the boundary between the EU and EU associate members.

Support for cross-border co-operation between ECE countries and EU Member States is provided respectively by the EU programme PHARE and INTERREG II. PHARE CBC is intended to support bi-lateral projects between Germany or Austria on the one hand, and the Czech Republic, Hungary or Poland on the other. There is also additional support for projects carried out in the border regions comprising areas of one EU Member State and two ECE countries, for example between Germany, the Czech Republic and Poland. PHARE CBC programme was introduced in 1994 and was available for five-year projects. For example, in the case of Czech-German co-operation, the Czech Republic received support amounting to ECU 125 million, 1995–99, while Czech-Austrian co-operation was supported by assistance to the ECE country amounting to ECU 30 million over the same period. PHARE CBC grants were especially provided for investment in transport, technical services and environmental infrastructure, and were also available for economic development, agriculture and labour policy development. To obtain a grant, the investor – which is usually a public sector institution – must cover at least 25 per cent of total costs. Support is given to mutually developed projects or to projects localised in the delimited frontier zone which will have an impact on both sides of the boundary.

Although the EU and national governments directed and supported cross-boundary co-operation, a number of other cross-border schemes emerged involving local government. Arguably the most notable is the German-Polish cross-border scheme. The German border area is peripheral both in relation to the heartlands of Germany and the EU as a whole. While it is situated in the poorest part of Germany (the *Länder* of Brandenberg and Mecklenburg-Vorpommern) it is relatively well developed when compared to the other side of the border. It has been suggested that if the cross-border region is to benefit from political, economic and social transformation, co-operation and the integration of development strategies are necessary (Boel, 1994). There are, however, major obstacles to co-operation. There are substantial attitudinal and linguistic differences between Germany and Poland, there are poor economic opportunities in the border region, and development (if it is to take place) will be within different political and economic contexts – Germany being a member of the EU for the time being and Poland not.

It was hoped that the introduction of the *Oderlandplan* in 1991 would ensure the co-operation of the relevant German and Polish local authorities – the thirty-four German *Kreise* and five Polish counties. On the German side of the border, the plan emphasised the need to stimulate agriculture in the north, high-tech industries in the south, the establishment of national parks and tourist areas, and the development of Frankfurt-an-der-Oder as an international transport focus, and economic and cultural centre. In the Polish counties, an essentially agricultural area would be diversified by the promotion of light industry in provincial capitals and a new north-side transport axis would be developed between Szczecin and central areas of eastern Europe. The *Oderlandplan* is essentially top-down rather than an attempt to develop an integrated top-down-bottom-up approach to development. It was consequently criticised by the CEC in 1994, and by 1997 it had still not qualifed for funding (Mainstone, 1997).

Since 1992, throughout much of East and Central Europe attention has focused on smaller scale co-operation (Bojar, 1996; Kortus, 1996) and several Euroregions have been established for this purpose (Table 12.10). In Hungary, in addition to the projects eligible for PHARE CBC assistance along the Austro-Hungarian border, Hungarian counties are involved in the Alps-Adria Work Community, the Danube Region Co-operation scheme, the Wien-Bratislava-Györ development triangle, the Carpathian Euroregion and the Alföod-Bánát-Vajdaság co-operation scheme. Within the Euro-regions, local authorities undertake bottom-up activity which often creates tension and conflict between national and local government, as this is also the case between the Romanian and Slovakian governments in the Carpathian Euroregion (Corrigan *et al.*, 1997). In the Czech Republic, the non-existence of regional self-government is seen as the major obstacle to more integrated co-operation within the Euroregions.

Table 12.10 Euroregions of East and Central Europe

Neisse-Nisa-Nysa Euroregion	PL, D, CZ
Pro Europa Viadrina Euroregion	PL, D
Pomerania Euroregion	PL, D, DK, SWE
Spree-Nysa-Bóbr Euroregion	PL, D
Upper Silesia-North Moravia Euroregion	PL, CZ
Carpathian Euroregion	PL, UKR, SK, H, R
Tatra Euroregion	PL, SK
Bayerischer Wald-Šumava	CZ, D
Egrensis	CZ, D
Erzgebirge-Krušné Hory	CZ, D
Elbe-Labe	CZ, D

The convergence of planning systems

Both within the EU and among its potential members, the extent to which planning systems vary provides a fruitful area for research. Despite different legal and governmental systems and varying degrees of decentralisation, there is undoubtedly some indication that spatial planning systems are beginning to converge as a result of the devolution of responsibility and the streamlining of planning procedures (CEC, 1994c). However, notwith-standing pressure from the European Parliament and the Committee of Regions for more integrated spatial planning, it is unlikely that there will be a complete harmonisation of planning systems throughout the EU in the foreseeable future – even within cross-border regions – because of the need for unanimous decision-making among Member States, an adherence to the principle of subsidiarity, and factors such as national attitudes, language and culture. Clearly, if cross-border co-operation and planning is to become a major element in the further development of the EU, these difficulties need to be overcome.

The polarisation of urban development

Although the basic structure of the urban hierarchy within the EU is likely to remain the same, it can be expected that further polarisation of urban growth will occur well into the twenty-first century. London and Paris and other cities at the top level of the hierarchy (including small towns and cities in their hinterlands) will continue to experience economic growth based on the development of a modern infrastructure and the exploitation of advanced technology and services (Wegener, 1995). In contrast, there will be a large number of potential losers, particularly those cities not linked to high-speed transport infrastructure, cities at the European periphery, or cities which have failed to rid themselves of old and inefficient industries. As with regional imbalance, the polarisation of urban development 'is in direct conflict with the stated equity goals of the regional policy of the European Union' (Wegener, 1995: 158).

Polarisation is not without its danger to cities experiencing economic growth. The indirect or social costs of growth include soaring land values, urban sprawl, traffic congestion and environmental degradation, while the very processes of urban development often ignore or intensify the needs of a growing number of socially excluded households particularly in the inner cities or peripheral estates.

The Single European Market will undoubtedly facilitate inter-regional exchange and trade, but might worsen the economic plight of cities on the outer extremities of the EU, particularly if they are not connected to new high-speed rail and telecommunications networks.

The need to reduce disparities between cities, no less than the need to reduce regional inequalities, is an overriding aim of regional policy. Indeed, since 80 per cent of the population of the EU is urban, this is clearly self-evident. However, while policy initiatives in Europe address many of the problems affecting European cities, 'these efforts have often been piecemeal, reactive and lacking in vision' (CEC, 1997c: 3). New efforts are undoubtedly necessary to strengthen or restore the role of Europe's cities as sources of economic prosperity and sustainable development, as places of social and cultural integration, and as bases for democracy (CEC, 1997c).

While the European Parliament and the Committee of Regions have supported greater intervention from the EU in urban development, and the European Commission and the Member States acknowledge their common concern about the future sustainability of cities, it is mainly within a regional context through the medium of Structural Funds that towns and cities have to compete for EU aid. With the reform of the Structural Funds for the programme period 2000–06, the distribution of aid will inevitably have a very varied effect on the urban areas of the EU and might do comparatively little to ameliorate the diseconomies of urban polarisation.

Structural reforms

Whereas, from their inception, it was intended that the Structural Funds should be concentrated in areas of greatest need, by the 1990s it was evident that resources were being spread too widely and too thinly. Objective 1, 2, 5b and 6 regions contained almost 51 per cent of the EU's population as a consequence of Objective 1 and Objective 6 regions being designated in thirteen of the fifteen EU Member States by the 1993 reforms. It was thus possible to make a strong case for reducing the geographical coverage of these regions to 35–40 per cent of the EU's population, for allocating aid more selectively in the cohesion countries and for targeting funds at development opportunities in areas of high unemployment in Objective 2 regions, rather than distributing resources widely across their length and breadth (Bachtler and Turrok, 1997). It has also been argued that there were too many funds and objectives. Within a complicated matrix of expenditure allocation, the allocation of structural policy resources from the ERDF, ESF, EAGGF, FIFG and Community Initiatives to Objective 1,2,3,4,5a,5b and 6 regions required a formidable bureaucratic structure at both the Commission and Member State level – made even more complex when regional programmes attempted to exploit the funding facilities of the European Coal and Steel Community and European Investment Bank (Bachtler and Turok, 1997).

In addition to reforms in these areas, it is probable that greater importance should be attached to the promotion of the small and medium-sized enterprise (SME) sector within the European economy. Since this sector is viewed as crucial in the creation of employment opportunities, greater emphasis should be given to the targeting of resources towards deprived urban areas in an attempt to overcome social exclusion, and that greater attention should be paid to the protection of the environment, for example by ensuring that Structural Funds are more precisely targeted at sustainable economic development.

Incorporating several of these proposals, *Agenda 2000* (CEC, 1997d) initiated the formal process of Structural Reform. Although economic and social cohesion had to remain a political priority, not least because of the impending enlargement of the Union, *Agenda 2000* proposed a ceiling on Structural and Cohesive Fund spending of 0.46 per cent of Union GDP, equal to a projected 275 billion ECU for the period 2000–06, of which 45 billion ECU would be allocated to potential Member States in East Central Europe to help them prepare for accession (Bachtler, 1998). However, the population coverage of the existing recipient areas would be cut from 51 per cent to 35–40 per cent of the total population of the EU, the seven Structural Fund Objectives would be reduced to three, and the thirteen Community Initiatives would similarly be restricted to three fields.

Agenda 2000 also proposed that Objective 1 designation would continue to apply to regions lagging behind in development and remain the most important focus of investment with the continued allocation of two-thirds of Structural Fund allocation. Programmes would emphasise the need to improve competitiveness through measures to improve the infrastructure, support innovation and SMEs, and develop human resources (Bachtler, 1998). While areas such as South Yorkshire, Merseyside, Sicilia, Kriti, Galicia and parts of eastern Germany would be eligible for Objective 1 funding if their GDPs per capita remained at less than 76 per cent of the EU average, transition periods might apply to regions losing eligibility (such as the Scottish Highlands and Islands and Northern Ireland), and special arrangements would be made for Objective 6 areas when they become re-designated as Objective 1 recipients.

Objective 2 designation (with former Objective 4–5 sub-elements) would apply to those regions suffering from economic and social restructuring problems and a high rate of unemployment or depopulation. These areas will be undergoing changes in their industrial or service sectors, and will contain urban areas in difficulty, rural areas in decline and crisis-hit areas dependent upon the fishing industry. New programmes would focus on diversification, measures to combat social exclusion, local development, increased support for SMEs, innovation, vocational training, and environmental protection (Bachtler, 1998). Clearly many old Objective 2 regions would retain their designation, although some areas such as central Scotland and Wales could face disqualification since their unemployment rate was well below the EU average in the late 1990s.

In regions not covered by Objectives 1 and 2 (largely the old Objective 3 and 4 areas), new Objective 3 regions would be designated where there was a perceived need to adapt and modernise education, training and employment systems. Funding would therefore facilitate associated economic and social change, lifelong education and training systems, active labour market policies to fight unemployment, and measures to deal with social exclusion.

Community Initiatives would be rationalised, and be focused entirely on:

i   cross-border, trans-national and inter-regional co-operation to promote harmonious and balanced spatial planning;
ii   rural development; and
iii human resources especially equal opportunities.

The terms of reference of other CIs could be incorporated in the Objective 1, 2 and 3 programmes, while the share of the Structural Fund budget allocated to the CIs would be reduced to 5 per cent (Bachtler, 1998). However, funds would be focused on considerably fewer but larger CIs during the programming period 2000–06 and the effects in any given area would be consequently more substantial than hitherto.

Agenda 2000 further proposed that the Cohesion Fund should be retained in its current form for Member States with GNPs per capita of less than 90 per cent of the EU average and participating in the third phase of EMU (Bachtler, 1998).

The proposals of *Agenda 2000* met with mixed responses from Member States, for example while there was widespread support for a rigid application of the criteria for designating Objective 1 regions, and a stoic acceptance that most of the Objective 1 areas in the wealthier EU countries (such as the UK) would be de-designated, there was much concern in Ireland and in the Lisboa region of Portugal that the loss of Objective 1 status would put at risk further economic convergence. Member States were also divided over proposals to concentrate CIs on a smaller number of larger programmes and over whether or not the Cohesion Fund should be retained in its current form (Bachtler, 1998). It must be borne in mind, however, that the contents of *Agenda 2000* were only proposals. These required further bilateral discussion between the EU and Member States and debate in the European Parliament before new Structural Fund Regulations could come into effect in the year 2000.

The European Monetary Union and regional development

The reform of the Structural Fund and changes in regional eligibility, however, might seem premature and ill-timed since the new programme period 2000–06 coincides with the early years of the European Monetary Union (EMU). Although it is argued that the introduction of a single currency and common interest rates across most of the EU in 1999 will consolidate the formation of a single market and bring certainty and security to business and help save or create jobs, it will also have differential impact on regional development and might destroy jobs. Whereas a single currency in the United States has resulted in labour migrating away from high to low wage-cost regions – with the effect of stabilising the level of unemployment between 1972–92, labour market flexibility in the EU was far less evident. When regions of Europe became uncompetitive during the same period, they did not adjust by exporting labour but by incurring job-loss – the level of unemployment in the EU as a whole trebling, (Stott, 1998). Also in contrast to the United States, whichever regions in the EU had the highest unemployment rates in 1983 they also had the highest percentages in 1993. Unemployment rates in Europe over the same period thus showed little sign of converging either with national means or the EU average (Martin, Tyler and Baddeley, 1998).

In focusing on regions in France, Germany, Italy, the Netherlands and the UK, Martin *et al.* (1998) ascribe differences in unemployment rates to major differences in the degree of labour market flexibility. In the older industrial areas and in some of the less developed regions, wages are much more rigid than elsewhere. In analysing the impact of EMU upon labour market flexibility, Martin *et al.* suggest that a rapid move towards EMU could nullify the effects of twenty years of European regional policy. Since EMU will preclude members from devaluing their currencies or lowering their interest rates when faced with uncompetitiveness in major sectors of their economies, they will have little option but to adjust to a substantial increase in unemployment in the most depressed regions (Stott, 1998). Unlike the American model, mass-migration will not provide a solution, since for cultural, social and linguistic reasons it would be unlikely for unemployed labour from, for example, Strathclyde or Newcastle to migrate in large numbers to, say, Milano or Stuttgart, while the other possibility – massive subsidies from richer to poorer regions – would be unacceptable to potential donors and thus be highly contentious politically.

Assuming that the UK joined the EMU at its start, and based on an analysis of labour market flexibility, 1976–94, Martin *et al.* (1998) suggest that the regions most affected by the conditions of EMU membership were, in order of vulnerability: Campania, Sud, Berlin, Northern Ireland, Sardegna, Sicilia, Abruzzo-Molise, Lazio, Scotland and the North West; and in order of those that will do best: East Anglia, Emilia-Romagna, East Midlands, Ile de France, Centro, Ouest, Sud-Ouest, Est, Centre-Est and Bassin Parisien (see Figure 12.2).

The implications of these findings suggest that there is a disproportionate need in the vulnerable regions to equip people at the bottom end of the labour market to obtain jobs. Although this must involve investment in appropriate education and education, it will also bring calls for a degree of 'downward flexibility', i.e. labour pricing itself into jobs through wage reductions. Governments of Member States and the EU will need to 'grasp the nettle' and ensure that there is an adequate flow of financial resources from richer to poorer regions to counteract the effects of EMU membership. It might be questioned whether Structural and Cohesive Fund spending, as programmed for 2000–06, will be sufficient to achieve transfers on a

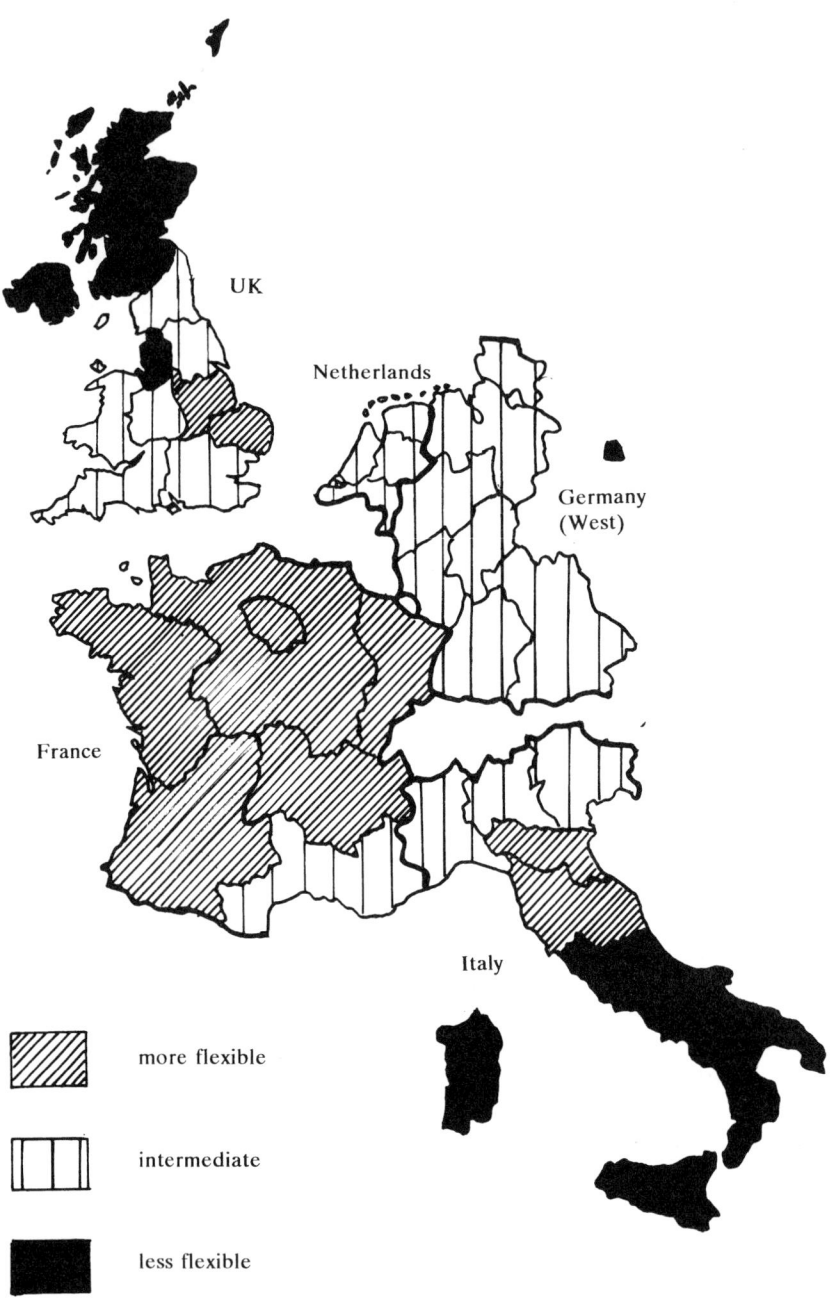

Figure 12.2 The predicted impact of Economic and Monetary Union on selected
regions in the EU with flexible and inflexible labour markets
Source: Derived from Martin *et al.*, 1998

sufficient scale, or targeted with sufficient precision, to compensate for the disadvantages of labour market inflexibility.

The enlargement of the European Union

The Czech Republic, Hungary and Poland are currently, in accordance with association agreements, in the process of adjusting their institutions and legislation to EU standards and are hoping to be in the next round of EU enlargement. The major obstacles on the road to the EU are: economic, technological and infrastructure backwardness; poor economic performance reflected in deviations from the Maastricht criteria, especially in the case of inflation; differences in legal systems, such as restrictions on real estate acquisition and limitations on economic activities by foreigners; non-existence of legal norms and institutions in several fields, such as regional government and regional development policies or the protection of intellectual property; and agricultural policy in the case of Poland (Blažek, 1997a; Szul and Mync, 1997; Gibb and Michalak, 1994). The main preconditions for the acceptation of ECE countries on the EU side is the willingness of Member States to support associated and later Member States from Structural Fund (SF), reform Common Agricultural Policy and prepare the adjustment of political representation in the basic EU institutions. It is expected, that there will be an adjustment period from 2000 to 2006 in which the new countries will adapt to the EU structural policies (Blažek, 1997a).

The association agreements with the EU have an important influence on the constitution of regional policies and transformation of physical planning in the Czech Republic, Hungary and Poland. The three countries will become eligible for financial assistance from the Structural Funds from 2000. Importantly, most of the territory of ECE countries will be eligible for Objective 1 status. For instance, the average GDP in the Czech Republic is 55 per cent of the EU average. However, the city of Praha's GDP is 82 per cent of the EU average and, if the city itself is considered a NUTS-2 region, it will probably be the only area in ECE countries that will not fall within the brackets of Objective 1 funding. Inevitably, with the inclusion of ECE countries the average GDP of the EU will drop and many areas formerly eligible for funding may no longer qualify.

There are several limitations for the utilisation of Structural Funds. The contemporary institutional arrangements and organisation of national regional policy does not, as yet, conform with EU requirements. The application within the Community Support Framework requires regional development plans to be elaborated and there are no regional governments in the Czech Republic and Poland. In addition, the Regional and County Development Councils in Hungary do not seem to fill appropriately the gap of a missing elected regional authority. It is also questionable whether local institutions in ECE countries will be able to adapt quickly to EU procedures, provide required detailed information about the projects and their implications for regional development and clearly specify financial arrangements. The co-financing of regional development projects is another very serious obstacle due to the scarce financial resources in ECE countries.

If the impact of EMU on EU regional policy is likely to be substantial, the effects of the enlargement of the Union to embrace ECE countries could be even more dramatic, though not necessarily so. While it is probable that, as a result of low cost labour-intensive production, there could be substantial trading advantages to the first wave of new Member States joining the EU around 2003–05, and also benefits to the border regions of the EU as a result of increased manufacturing output and employment, the advantages to Western Europe could be negligible as marginal sectors such as agriculture, steel and textiles might, in part, be rendered uncompetitive. Clearly, the West could benefit from the export of capital goods for the purpose of restructuring, but it could lose by opening up its markets to imports of agricultural produce and low-tech, labour-intensive products which would undermine the poorest EU regions such as areas in Greece and Portugal (Hall and Van der Wee, 1995) or put at risk the competitive ability of peripheral regions in which there is a large or above average share of agricultural or manufacturing employment, for example in Orense, la Corruna, Almeria and Burgos (Bachtler and Turok, 1997). In contrast to Austria, Finland and Sweden (which joined the EU in 1995), the new Member States will not be net contributors to the EU budget. Whereas Greece (the poorest EU member) had a GDP per head of \$11,500 in 1996, in the same year the comparable figures for Slovenia were \$9279, \$4357 Hungary, \$5340 the Czech Republic and Poland \$3459 (*Economist*, 1996; European Bank for Reconstruction and Development, 1997). This would suggest that cohesion policies will be skewed towards the new Member States if convergence is to remain the principal objective of EU regional policy (Bachtler and Turok, 1997).

In negotiating an enlargement of the EU, there is undoubtedly a conflict of national interests. Clearly there is a need to expand the Structural Fund Budget, but this is opposed by most of the net contributors who would prefer to reduce rather than increase payments to the EU budget. They clearly fear that 'regional policy within the framework of a larger EU would primarily be a device for channelling funding eastward. Nearly all the existing fifteen Member States would find themselves becoming net contributors (Fothergill, 1997). It would also be necessary for those Members that were major recipients of Structural Fund support in 1993–99 to forgo some or all of their allocations, but this is strongly opposed by the Members concerned since their ability to compete in Europe would be impaired by the loss of regional aid; and while there may be a need to treat new Member States differently, this is opposed by ECE countries since, for reasons of economic and political status, they seek equality of treatment with existing Members (Bachtler and Turok, 1997).

At the centre of the debate is the question of cost. Recent research suggests that if the EU were enlarged to include the Czech Republic, Hungary, Poland and Slovakia, the Structural Fund budget would need to increase by 12 to 26 billion ECU annually (it is probable that broadly the same budgetary increase would be necessary if Slovenia rather than Slovakia were to accede in the first wave of new Members).

However, the problems of enlargement, discussed above could be mitigated. Economic growth rates in Central and Eastern Europe, already higher than the EU average in the mid-1990s, might remain high well into the first decade of the twenty-first century, and several cities and regions consequently might exceed the eligibility ceiling for Objective 1 status, for example Praha. But if the GDP per capita of most regions remained low, it is probable that large scale transfers would either be unnecessary since, by EU standards, small amounts of support would have a major impact on investment and purchasing power in the recipient countries, or, the recipient countries would not have the institutional, financial and economic capacities to effectively absorb major Structural or Cohesive Fund transfers (Bachtler and Turok, 1997). As a consequence of these mitigations, the costs of EU enlargement could be considerably lower than that predicted, although existing recipients would undoubtedly suffer some reduction in support. Since it is unlikely that enlargement will take place much before 2006, the EU can continue until that time to seek advice from national, regional and sectoral interests, propose ways in which the beneficial effects of enlargement can be exploited and the adverse effects of enlargement ameliorated, and engage in constructive discussion with both existing and prospective Members before a definitive programme is agreed.

References

Abercrombie, P. and Matthew, R. (1946) *The Clyde Valley Regional Plan 1946*, Edinburgh: HMSO.

Acosta, R. and Renard, V. (1991) *Frameworks and Functioning of Urban Land and Property Markets in France*, Paris: Association des Etudes Foncières.

Act no. 50/1976 on Physical Planning and the Building Act (the Czech Republic).

Act XXI of 1996 (IV. 5.) on Regional Development and Physical Planning (Hungary).

Allen, K. and MacLennan, M. (1970) *Regional Problems and Policies in Italy and France*, London: Allen & Unwin.

Alm, J. and Buckley, R.M. (1994) 'Decentralization, privatization, and the solvency of local governments in reforming economies: the case of Budapest', *Environment and Planning C: Government and Policy* 12, 4: 333–346.

Armstrong, H. and Taylor, J. (1993) *Regional Economies and Policy*, 2nd Edition, St Albans: Harvester Wheatsheaf.

Assembly of Welsh Counties (1992) *Strategic Planning Guidance in Wales – Overview Report*, Mold: Clwyd County Council.

Ave, G. (1991) *Urban Land and Property Markets in Italy*, London: UCL Press.

Bachtler, J. (1998) '*Agenda 2000: Implications for the Structural Funds Regions*', Regional Studies Association no. 213.

Bachtler, J. and Turrok, I. (1997) *The Coherence of EU Regional Policy*, Jessica Kingsley Publishers: London.

Balchin, P. (1979) *Housing Improvement and Social Inequality*, Farnborough: Saxon House.

Balchin, P.N. and Bull, G.H. (1987) *Regional and Urban Economics*, London: Harper Row.

Bangemann, M. (1994) *Europe and the Global Information Society*, DG XIII, Brussels.

Barlow Report (1940) *Report of the Royal Commission on the Distribution of the Industrial Population*, cmd 6153, London: HMSO.

Bassols, M. (1986) 'Town planning in Spain' in Garner, J.F. and Gravells, N.P. (eds) *Planning law in Western Europe*, Amsterdam: Elsevier Science Publishers.

Bennett, R.J. (ed.) (1993) *Local Government in the New Europe*, London: Belhaven Press.

Biarez, S. (1993) 'Ville, région, état, le dialogue en Europe', in J.-C. Némery and Wachter, S. (eds) *Entre L'Europe et la décentralisation*, La Tour d'Aigues: DATAR/ Editions de l'Aube.

Bird, R.M. (1994) 'Local finance and economic reform in Eastern Europe', *Environment and Planning C: Government and Policy* 12, 3: 263–276.

Blažek, J. (1994) 'Changing local government finances in the Czech Repubic – half way over?', *GeoJournal* 32, 3: 26–267.

Blažek, J. (1996) 'Regional patterns of adaptability to global and transformation processes in the Czech Republic', *Acta Facultatis Rereum Naturalium Universitatis Comenianae – Geographica*, 37: 61–70.

Blažek, J. (1997a) 'The Czech Republic on its way towards the West European structures', *European Spatial Research and Policy* 4, 1: 37–62.

Blažek, J. (1997b) 'Orientation paper on regional policy in the Czech Republic for PAU of the Ministry of Regional Development', manuscript.

Blažek, J. and Kára, J. (1992) 'Regional policy in the Czech Republic in the period of transition', in G. Gorzelak and A. Kukliñski (eds) *Dilemmas of Regional Policies in Eastern and Central Europe*, Warsaw: University of Warsaw, 7–94.

Blažek, J., Hampl, M. and Sýkora, L. (1994) 'Administrative system and development of Prague', in M. Barlow, P. Dostál and M. Hampl (eds) *Development and Administration of Prague*, Amsterdam: Universiteit van Amsterdam, 73–87.

Blacksell, M. and Williams, A.M. (1994) *The European Challenge*, Oxford: Oxford University Press.

Blotevogel, B.H. and Fielding, A.J. (1996) *People Mobility and Jobs in the New Europe*, Chichester: Wiley.

BM Bau (1993) *Guidelines for Regional Planning: General Principles for Spatial Development Planning in the Federal Republic of Germany*, Bonn: Bundesbauministerium.

Boel, S. van de (1994) 'The challenge to develop a border region: German-Polish cooperation', *European Spatial Research and Policy* 1, 1: 57–72.

Bojar, E. (1996) 'Euroregions in Poland', *TESG – Tijdschrift voor economische en sociale geografie* 87, 5: 442–447.

Brotchie, J., Batty, M., Blakely, E., Hall, P. and Newton, P. (1995) *Cities in Competition: Productive and Sustainable Cities for the 21st Century*, Melbourne, Australia: Longman.

Brussard, W. (1986) 'Physical planning legislation in the Netherlands' in Garner, J.F. and Gravells, N.P. (eds) *Planning Law in Western Europe*, Amsterdam: Elsevier Science Publishers.

Bull, G. H. (1996) 'Social Polarisation and Housing in Europe', paper presented at ENHR Conference, Copenhagen, August.

Bullmann, U. (1994) 'Regionen im Integrationsprozess der Europaischen Integration,' in Bullmann, U. (ed.) *Die politik der dritten ebene. regionen im Europa der union*, Baden-Baden: nomos.

Bullman, U. (1997) 'The politics of the thir level', in C. Jeffery (1997) *The Regional Dimension of the European Union*m London: Frank Cass.

Burtenshaw, D., Bateman, M. and Ashworth, G.J. (1991) *The European City: a Western Perspective*, London: David Fulton.

Carter, F.W. (1979) 'Prague and Sofia: an analysis of their changing internal city structure', in R.A. French, F.E.I. Hamilton (eds) *The Socialist City: Spatial Structure and Urban Policy*, Chichester: John Wiley & Sons, 425–459.

CEC (1985) *Completing the Internal Market*, White Paper, Brussels: CEC.

CEC (1987) Third Periodic Report on the Social and Economic Situation and Development of the Regions of the Community, Brussels: OOPEC.

CEC (1988a) Research and Technological Development Policy, European Documentation 2/1988, Luxembourg: CEC.

CEC (1988b) 'The Economics of 1992: An Assessment of the Potential Economic

Effects of Completing the Internal Market of the European Community', European Economy, No. 35, Luxembourg: OOPEC.

CEC (1990a) *The Impact of the Internal Market by Industrial Sector*, Commission of the EC, European Economy, Special Edition, OOPEC, Luxembourg.

CEC (1990b) Greenbook on the Urban Environment (Com (90) 218) Brussels.

CEC (1991a) Regions in the 1990s, Fourth Periodic Report, Brussels: European Commission.

CEC (1991b) *Europe 2000*, Brussels: Commission of the EC.

CEC (1992a) Fifth action programme on the environment 1992–2000: towards sustainability (com (92) 23) Brussels.

CEC (1992b) *Regional Development Studies: Inter-regional and Cross-border Co-operation in Europe*, Brussels: European Commission.

CEC (1993a) *'Community Public Finance in the Perspective of EMU'*, European Economy, 53.

CEC (1993b) *Growth, Competitiveness and Employment*, White Paper, Brussels: Commission of the EC.

CEC (1993c) *Communication from the Commission to the Council, the European Parliament and the Economic and Social Committee on Energy and Economic and Social Cohesion*, Brussels: European Commission.

CEC (1994a) *Report on Science and Technology Indicators*, Luxembourg: OOPEC.

CEC (1994b) *Competition and Integration: Community Merger Control Policy Commission of the EC*, European Economy No. 57, Luxembourg: OOPEC.

CEC (1994c) *Europe 2000+: Cooperation for European Territorial Development*, DG XVI, Brussels.

CEC (1994d) *Fifth Periodic Report on the Social and Economic Situation and Development of the Regions of the Community*, Brussels.

CEC (1996a) *Report on Economic Cohesion*, Brussels: CEC.

CEC (1996b) *'Europe at the Service of Regional Development'*, Brussels: CEC.

CEC (1997a) *'International Trade and Foreign Direct Investment 1996'*, Brussels: CEC.

CEC (1997b) *Towards Sustainability*; The European Commission's Progress Report and Action Plan on the Fifth Programme of Policy and Action in Relation to the Environment and Sustainable Development, Luxembourg: Office for Official Publications of the EC.

CEC (1997c) *Towards an Urban Agenda in the EU* (com(97) 197), Brussels: CEC.

CEC (1997d) *Agenda 2000*, Brussels: CEC.

CEC (1998a) *European Spatial Development Perspective*, Brussels: CEC.

CEC (1998b) *Compendium of European Spatial Planning*, Brussels: CEC.

Cecchini, P. (1988) *The European Challenge: 1992 The Benefits of a Single Market*. Aldershot: Wildwood House.

Champion, T., Mønnesland, J. and Vandermotten, C. (1996) *'The New Regional Map of Europe'*, Progress in Planning, 36: 1–89.

Cheshire, P. and Hay, A. (1989) *Urban Problems in Western Europe*, London: Allen and Unwin.

Chicoye, C. (1992) 'Regional impact of the Single European Market in France', *Regional Studies* 26,4.

Ciechocinska, M. (1994) 'Paradoxes of decentralisation: Polish lessons during the transition to a market economy', in R. Bennett (ed.) *Local Government and Market Decentralisation*, Tokyo: United Nations University Press.

Clout, H., Blacksell, M., King, R. and Pinder, D. (1995) *Western Europe: geographical perspectives*, London: Longman.

Cole, G.D.H. (1921) *The Future of Local Government*, London: Cassell.

Cole, J. and Cole, F. (1997) Geography of the European Union, 2nd edition London: Routledge.

Collins, K. and Schmenner, R. (1995) 'Taking Advantage of Europe's Single Market', *European Management Journal* 13: 257–268.

Committee for the Study of Economic and Monetary Union (1989) *Report on the EMU in the EC*, Luxembourg: OOPEC.

Corrigan, J., Süli-Zakar, I. and Béres, C. (1997) 'The Carpathian Euroregion: an example of cross-border co-operaton', *European Spatial Research and Policy* 4, 1: 113–124.

Cullingworth, J.B. and Nadin, V. (1994) *Town and Country Planning in Britain*, 11th edition, London: Routledge.

Cullingworth, J.B. and Nadin, V. (1997) *Town and Country Planning in the UK* (twelfth edition), London: Routledge.

Davies, H.W.E. (1989) *Planning Control in Western Europe*, London: HMSO.

Delladetsima, O. and Leontidou, L. (1995) 'Athens' in J. Berry and S. McGreal (eds) *European Cities: Planning Systems and Property Markets*, London: E. & F.N. Spon.

Department of Trade and Industry (1983) *Regional Industrial Development*, cmnd 9111 London: HMSO.

Dixon, R.J. and Thirlwall, A.P. (1979) 'A model of growth rate differentials along Kaldorian lines', *Economic Journal*, 89.

Doe/Welsh Office (1986) 'The future of development plans: a consultation paper', London: DOE.

Dostál, P., Illner, M., Kára J., and Barlow, M. (eds) (1992) *Changing Territorial Administration in Czechoslovakia: International Viewpoints*, Amsterdam: Universiteit van Amsterdam, 17–32.

Drake, G. (1994) *Issues in the New Europe*, London: Hodder and Stoughton.

Douglas, M.J. (1997) *A Change of System: Housing System Transformation and Neighbourhood Change in Budapest*, Utrecht: Urban Research Centre.

Duncan, S. (1985) 'Land policy in Sweden: separating ownership from development', in S. Barrett and P. Healey (eds) *Land Policy: Problems and Alternatives*, Aldershot: Avebury.

Eaton, M. and Lipsey, R. (1979) 'The Theory of Market Pre-emption: the presence of excess capacity and monopoly in growing spatial markets', *Economics*, vol. 46.

Eatwell, J. (1995) 'The International Origins of Unemployment' in J. Michie and J. Smith (eds) *Managing the Global Economy*, Oxford: Oxford University Press.

EC Background Report 8/3/97

Economist, The (1996) *The World in 1996*, London.

Enyedi, G. (1990a) *New Basis for Regional and Urban Policies in East-Central Europe*, Pécs: Centre for Regional Studies of Hungarian Academy of Sciences.

Enyedi, G. (1990b) 'Private economic activity and regional development in Hungary', *Geographia Polonica* 57: 53–62.

Enyedi, G. (1994) 'The transition of post-socialist cities', *European Review* 3, 2: 171–182.

EP (1997) *European Parliament News*, June 1997.

European Bank for Reconstruction and Development (1997) *Transition Report*, London: EBRD.

Eurostat (1997) *Europe in figures*, 6th edition, Luxembourg: Office for official publications of the EC.

Eversley, D. (1972) 'Rising Costs and Static Incomes: some consequences of regional planning in London', *Urban Studies*, vol. 9.

Fairclough, A.J. (1983) 'The Community's environmental policy' in R. MacRory, (ed.) *Britain, Europe and the environment*, London: Imperial College.

Falkanger, T. (1986) 'Planning law in Norway', in J.F. Garner and N.P. Gravells (eds) *Planning Law in Western Europe*, Amsterdam: Elsevier Science Publishers.

Faragó, L. (1994) 'The competitive development concept of the South Trans-danubian region', in Z. Hajdú and G. Horváth (eds) *European Challenges and Hungarian Responses to Regional Policy*, Pécs: Centre for Regional Studies.

Fawcett, C.B. (1917) *The Provinces of England*, London: Hutchinson.

Fielding, A., (1989) 'Migration and Urbanisation in Western Europe Since 1950', *The Geographical Journal*, 155.

Fothergill, S. (1997) Address to the conference of the Regional Studies Association, Frankfurt-an-der Oder, 20–23 September.

Geddes, P. (1915) *Cities in Evolution*, London: Ernest Benn.

Getimis, P. (1992) 'Social conflicts and the limits of urban policies in Greece', in M. Dunford and G. Kafkalas (eds) *Cities and Regions in the New Europe*, London: Belhaven Press.

Gibb, R.A. and Michalak, W.Z. (1994) 'The European Community and East-Central Europe', *TESG – Tijdschrift voor economische en sociale geografie* 85, 5: 401–416.

Gorzelak, G. (1996) *The Regional Dimension of Transformation in Central Europe*, Regional Policy and Development Series 10, London: Regional Studies Association and Jessica Kingsley Publishers.

Gripaios, P. (1995) 'The Role of European Super-Regions', *European Urban and Regional Studies*, 1995, 2.1: 77–81.

Gripaios, P. and Mangles, T. (1993) 'An analysis of European Super Regions' *Regional Studies* 27, 8.

Grochowski, S. (1997) 'Public administration reform: an incentive for local trans-formation?', *Environment and Planning C: Government and Policy* 15, 2: 209–218.

Group of Focal Points (1994) *Vision and strategies around the Baltic Sea 2010: towards a framework for spatial development in the Baltic Sea region*, Karlskrona: The Baltic Institute.

Grubel, H. (1967) 'Intra-Industry Specialisation and the Pattern of Trade' *Canadian Journal of Economics and Political Science*, vol. 22.

Guichard, O. (1986) *Propositions pour l'aménagement du territoire*, Paris: La Ducumen-tation Française.

Hajdú, Z. (1993) 'Local government reform in Hungary', in R.J. Bennett (ed.) *Local government in the new Europe*, London and New York: Belhaven Press, 208–224.

Hall, P. (1992) *Urban and Regional Planning* (third edition), London: Routledge.

Hall, R. and Van der Wee, M. (1995) 'The regions in an enlarged Europe', in S. Hardy, M. Hart, L. Albrechts, and A. Katos, (eds) *An Enlarged Europe: Regions in Competition*, London: Jessica Kingsley Publishers.

Hall, T. (ed.) *Planning and Urban Growth in Nordic Countries*, London: E. & F.N. Spon.

Hammersley, R. (1997) 'Environment and regenerations in the Czech Republic', *International Planning Studies* 2, 1: 103–122.

Hammersley, R. and Westlake, T. (1994) 'Urban heritage in the Czech Republic', in G.J. Ashworth and P.J. Larkham (eds) *Building a New Heritage: Tourism, Culture and Identity in the New Europe*, London: Routledge, 178–200.

Hardy, S., Hart, M., Albrechts, L. and Katos, A. (1996) *An Enlarged Europe: Regions in Competition*, London: Jessica Kingsley Publishers.

Harrop, J. (1996) *Structural Funding and Employment in the European Union*, London: Edward Elgar.

Healy, A. (1997) 'Europe needs you', *Planning*, 12 October.

Hill, S. and Munday, M. (1992) 'The UK Regional Distribution of FDI', *Regional Studies* 26.6.

Hirschman, A. (1958) *'The Strategy of Economic Development'*, New Haven, CT: Yale University Press.

Hitchens, D. (1997) 'Environmental policy and the implications for competitiveness in the regions of the EU', *Regional Studies*, 31, 8.

HM Government (1985) *Lifting the Burden*, London: HMSO.

HM Government (1989) *The Future of Development Plans*, cmnd 569, London: HMSO.

Hoffman, L.M. (1994) 'After the fall: crisis and renewal in urban planning in the Czech Republic', *International Journal of Urban and Regional Research* 18, 4: 691–702.

Holt-Jansen, A. (1990) 'Planning in Norway 1965–1990: changing philosophies, planning laws and practical results', Paper to AESOP Congress, Reggio Calabria, November.

Horváth, G. (1996) 'The regional policy of the transition in Hungary', *European Spatial Research and Policy* 3, 2: 39–55.

Horváth, T. M. (1997) 'Decentralization in public administration and provision of services: an East-Central European view', *Environment and Planning C: Government and Policy* 15, 2: 161–175.

IAURIF (1991) Institut d'Aménagement et d'Urbanism de la Région d'Ile-de-France, *La Charte de L'Isle-de-France*, Paris: IAURIF.

Jeffery, C. (ed.) (1997) *The Regional Dimension of the European Union*, London: Frank Cass.

Jones, R. (1996) *'The Politics and Economics of the European Union'*, London: Edward Edgar.

Judge, E. (1995) 'Warsaw', in J. Berry and S. McGreal (eds) *European Cities, Planning Systems and Property Markets*, London: E & FN Spon, 345–370.

Kalbro, T. and Mattison, H. (1995) *Urban Land and Property Markets in Sweden*, London: UCL Press.

Kaldor, N. (1971) 'The Dynamic Effects of the Common Market', in D. Evans (ed.) *Destiny or Delusion: Britain and the Common Market*, London: Gollancz.

Kára, J. (1992) 'Prague, the city growth and its administration', in P. Dostál, M. Illner, J. Kára and M. Barlow (eds) *Changing Territorial Administration in Czechoslovakia: International Viewpoints*, Amsterdam: Universiteit van Amsterdam, 33–38.

Kára, J. (1994) 'New Czech regional policy', in M. Barlow, P. Dostál and M. Hampl (eds) *Territory, Society and Administration: The Czech Republic and the Industrial Region of Liberec*, Amsterdam: Universiteit van Amsterdam, 67–83.

Kára, J. and Blažek, J. (1993) 'Czechoslovakia: regional and local government reform since 1989', in R.J. Bennett (ed.) *Local government in the New Europe*, London and New York: Belhaven Press, 246–258.

Keyes, J., Munt, I. and Riera, P. (1991) *Land Use Planning and the Control of Development in Spain*, Reading: University of Reading, Centre for European Property Research.

Kilbrandon Report (1973) *Report of the Royal Commission on the Constitution*, cmnd 5640, London: HMSO.

Kinnock, N. (1998) '*Europe on the Move: the Transport Agenda, Britain in Europe*, The European Institute: London.

Kivell, P. (1993) *Land and the City*, London: Routledge.

Kortus, B. (1996) 'Recent economic transformation in Poland', in F. W. Carter, P. Jordan and V. Rey (eds) *Central Europe after the Fall of the Iron Curtain: Geopolitical Perspectives, Spatial Patterns and Trends*, Frankfurt am Main: Peter Lang, 217–236.

Kunzmann, K.R. and Wegener, M. (1991) 'Pattern of urbanisation in Western Europe 1960–1990,' *Berichte aus dem institut fur raumplannung 28* Dortmund: universitat Dortmund.

Labour Party (1997) *A New Voice for England's Regions: Labour's Proposals for English Regional Government*, London: Labour Party.

Lackó, L. (1994) 'Settlement development processes and policies in Hungary', in Z. Hajdú and G. Horváth (eds) *European Challenges and Hungarian Responses to Regional Policy*, Pécs: Centre for Regional Studies.

Lawrence, R.J. (1996) 'Switzerland', in Paul Balchin (ed.) *Housing Policy in Europe*, London: Routledge.

Leemans, A.F. (1970) *Changing Patterns of Local Government*, The Hague: International Union of Local Authorities.

Le Galès, P. and John, P. (1997) 'Is the grass greener on the other side? What went wrong with the French regions, and the implications for England', *Policy and Politics* 25, 1: 51–60.

Le Monde (1997) *Régions*, F. Grosrichard (ed.), 16 Avril, 1997.

Lengyel, I. (1993) 'Development of local government finance in Hungary', in R.J. Bennett (ed.) *Local government in the new Europe*, London and New York: Belhaven Press, 225–245.

Lichtenberger, E. (1996) 'The geography of transition in East-Central Europe: society and settlement systems', in F.W. Carter, P. Jordan and V. Rey (eds) *Central Europe after the Fall of the Iron Curtain: Geopolitical Perspectives, Spatial Patterns and Trends*, Frankfurt am Main: Peter Lang, 137–152.

Lloyd, T.A. and Jackson, H. (1949) *Outline Plan for South Wales and Monmouthshire* London: HMSO.

Lorange, E. and Myhre, J.E. (1991) 'Urban planning in Norway', in T. Hall (ed.) *Planning and Urban Growth in the Nordic Countries*, London: E. & F.N. Spon.

Lorenzen, A. (1996) 'Regional development and institutions in Hungary: past, present and future development', *European Planning Studies* 4, 3: 259–277.

Loughlin, J. (1997) 'Representing regions in Europe: the committee of the regions' in C. Jeffery (ed.) *The Regional Dimension of the European Union. Towards a Third Level in Europe*, London: Frank Cass.

Lovering, J. (1997) 'Global Restructuring and Local Impact' in M. Pacione (ed.), *Britain's Cities: Geographies of Division in Urban Britain*, London: Routledge.

Luttrell, W.F. (1987) 'A programme for regional development: what could be done in five years, paper presented at the TCPA Annual Conference, Sheffield.

Mackintosh, J.P. (1968) *The Devolution of Power*, Harmondsworth: Penguin Books.

Maclennan, D. and Stephens, M. (1997) *'EMU and the UK Housing and Mortgage Markets'*, London: Council of Mortgage Lenders.

Maclennan, D. (1997) 'Britain's cities – new thoughts, new start?', *Town and Country Planning*, October.

Mainstone, S. (1997) Cross-border planning in Europe (unpublished MSc dissertation, University of Greenwich).

Mansikka, M. and Rausti, J. (1992) 'Finland', in A. Dal Cin and D. Lyddon (eds) *International Manual of Planning Practice* (second edition), The Hague: ISOCARP.

Marcou, G, and Verebelyi, I. (eds) (1993) *New Trends in Local Government in Western and Eastern Europe*, Brussels: International Institute off Administrative Sciences.

Martin, R. and Tyler, P. (1992) 'The regional legacy', in J. Michie (ed.) *The Economic Legacy 1979–1992*, London: Academic Press.

Martin, R., Tyler, P. and Baddely, M. (1997) 'If jobs matter, softly, softly is the right way to go for EMU' *The Observer*, 16 November.

Massey, D. (1989) 'Regional planning 1909–1939: the experimental era', in P.L. Garside, and M. Hebbert (eds) *British Regionalism 1900–2000*, London: Mansell.

Mazza, L. (1991) 'European viewpoint: a new status for Italian metropolitan areas', *Town Planning Review*, 62(2).

Menanteau, J. (1997) 'Regions', *Le Monde*, 13 February 1997.

MERP (1996) *National Regional Development Concept (Extract), Draft*, Budapest: Ministry of Environment and Regional Policy.

Middlemas, K. (1995) *Orchestrating Europe: the Informal Politics of the European Union*, London: Fontana Press.

Minford, P. (1997) 'Why Not a Single Currency?' Economic Review, April.

Ministry of Health (1921) 'Report of the South Wales Regional Survey Committee', London: HMSO.

Ministry of Housing and Local Government (1963) *The North-East: a Programme for Development and Growth*, cmnd 2206, London: HMSO.

Ministry of Housing and Local Government (1964) *South East England*, cmnd 2308, London, HMSO.

Minshull, G.N. (1996) *The New Europe in the 21st Century*, revised and updated by M.J. Dawson, London: Hodder & Stoughton.

Morphet, J. (1997) 'There'll be planning, but not as we know it', *Town and Country Planning*, April.

Musil, J. and Ryšavý, Z. (1983) 'Urban and regional processes under capitalism and socialism: a case study from Czechoslovakia', *International Journal of Urban and Regional Research* 7, 4: 495–527.

Myrdal, G. (1957) *Economic Theory and Underdeveloped Regions*, London: Duckworth

Nadin, V. (1998) 'Planning and the UK Presidency', *Town and Country Planning*, vol. 67(2).

Newman, P. and Thornley, A. (1996) *Urban Planning in Europe. International Competition, National Systems and Planning Projects*, London: Routledge.

Norton, A. (1983) *The Government and Administration of Metropolitan Areas in Western Democracies: Survey of Approaches to the Administrative Problems of Major Conurbations in Europe and Canada*, Birmingham: Birmingham University, Institute of Local Government Studies.

Nugent, N. (1994) *The Government and Politics of the European Union*, 3rd edition, Basingstoke: Macmillan.

O'Farrell, P.N., Wood, P.A. and Zheng, J. (1996) 'Internationalisation of Business Services' *Regional Studies*, 30.2.

Office for National Statistics (1997) Regional Trends, 32, London: The Stationery Office.

Östergard, N. (1994) *Spatial Planning in Demark*, Copenhagen: Ministry of the Environment.

Palard, J. (1993) 'Structural and regional planning confronted with decentralisation and European integration', *Regional Politics and Policy*, 3, 3, Autumn.

Pálné Kovács, I. (1993) 'The current problems of local/regional government in Hungary', in Z. Hajdú (ed.) *Hungary: Society, State, Economy and Regional Structure in Transition*, Pécs: Centre for Regional Studies, 55–67.

Paul, L. (1995) 'Regional development in Central and Eastern Europe: the role of inherited structures, external forces and local initiatives', *European Spatial Research and Policy* 2, 2: 19–41.

Pavlínek, P. (1992) 'Regional transformation in Czechoslovakia: towards a market economy', *TESG – Tijdschrift voor economische en sociale geografie* 83, 5: 361–371.

Perlín, R. (1996) 'Local Government – Process of Rebuilding in the Czech Republic', *Acta Facultatis Rereum Naturalium Universitatis Comenianae – Geographica*, 37: 141–147.

Phelps, N.A. (1997) *'Multinationals and European Integration'*, London: Jessica Kingsley Press for the Regional Studies Association.

Pinder, D. (1983) *Regional Economic Development: Theory and Practice in the European Community*, London: Allen and Unwin.

Population Reference Bureau (1995) World Population Data Sheet, Washington D.C.

Ratcliffe, J. and Stubbs, M. (1996) *Urban Planning and Real Estate Development*, London: UCL Press Ltd.

Redcliffe-Maud Report (1969) *Report of the Royal Commission on Local Government in England*, cmnd 4040, London: HMSO.

Regulska, J. (1987) 'Urban development under socialism: the Polish experience', *Urban Geography* 8, 4: 321–339.

Regulska, J. (1997) 'Decentralization or (re)centralization: struggle for political power in Poland', *Environment and Planning C: Government and Policy* 15, 2: 187–207.

Ringli, H. (1992) 'Switzerland', in A. Dal Cin and D. Lyddon (eds) *International Manual of Planning Practice* (second edition), The Hague: ISOCARP.

Ringli, H. (1994) 'Tendencies in development plan-making in Switzerland', in P. Healey (ed.) *Trends in Development Plan-making in European Planning Systems*, Working Paper No. 42, Department of Town and Country Planning, University of Newcastle upon Tyne.

Royal Commission (1969a) *Report of the Royal Commission on Local Government in England 1966–1969*, cmnd 4140, London: HMSO.

Royal Commission (1969b) *Report of the Royal Commission on Local Government in Scotland 1966–1969*, cmnd 4150, Edinburgh: HMSO.

Royal Commission (1973) Report of the Royal Commission on the constitution, cmnd 5460, London: HMSO.

Rubenstein, J.M. and Unger, B.L. (1992) 'Planning after the fall of Communism in Czechoslovakia', *Focus* 42, 4: 1–6.

Sanders, W. (1905) *Municipalization by Provinces*, London: Fabian Society.

Scargill, I. (1996) 'French planning rediscovers the pays', *Town and Country Planning*, January.

Scottish Development Department (1963) *Central Scotland: a Programme for Development and Growth*, cmnd 2188, Edinburgh, HMSO.

Scottish Office (1991a) Review of planning guidance – a consultation paper, Edinburgh: environment department.

Scottish Office (1991b) *The Structure of Local Government in Scotland*, Edinburgh: HMSO.

Scottish Office (1992a) *The Structure of Local Government in Scotland: Shaping the Future – the new councils*, Edinburgh: HMSO.

Scottish Office (1992b) Planning advice note 37, Glasgow: Structure Planning.

Scottish Office (1997) *A Government for Scotland*, Edinburgh: HMSO.

Sheehan, M. (1993) 'Government Financial Assistance and Manufacturing Investment in Northern Ireland', *Regional Studies*, 27.6: 527–540.

Smith, A. (1997) 'The French case: the exception to the rule?' in C. Jeffery, (ed.) *The Regional Dimension of the European Union*, London: Frank Cass.

Sotarauta, M. (1994) 'Finnish municipalities and planning in transition: empowerment, impulses and strategies', *European Planning Studies* 2, 3.

Stoker, G., Hogwood, B. and Bullmann, U. (1995) *Regionalism*. Report to the local government Management Board, Glasgow: University of Strathclyde.

Stone, I. and Peck, F. (1996) 'The Foreign Owned Manufacturing Sector in UK Peripheral Regions 1978–93', *Regional Studies*, 30.1

Stott, M. (1998) 'A warning for the regions', *Town and Country Planning*, January/February.

Strong, A.L., Reiner, T.A. and Szyrmer, J. (1996) *Transformations in Land and Housing: Bulgaria, The Czech Republic and Poland*, New York: St. Martin's Press.

Suetens, L.P. (1986) 'Town and country planning law in Belgium' in J.F. Garner and N.P. Gravells, (eds) *Planning Law in Western Europe*, Amsterdam: Elsevier Science Publishers.

Surazska, W. (1996) 'Transition to democracy and the fragmentation of a city: four cases of Central European capitals', *Political Geography* 15, 5: 365–381.

Surazska, W. and Blažek, J. (1996) 'Municipal budgets in Poland and the Czech Republic in the third year of reform', *Environment and Planning C: Government and Policy* 14, 1: 3–23.

Surazska, W., Bucek, J., Malikova, L. and Danek, P. (1997) 'Towards regional government in Central Europe: territorial restructuring of postcommunist regimes', *Environment and Planning C: Government and Policy* 15, 4: 437–462.

Swann, D. (1996) *European Economic Integration*, Edward Elgar, London.

Swianiewicz, P. (1992) 'The Polish experience of local democracy: is progress being made?', *Policy and Politics*, 20(2).

Sýkora, L. (1994a) *The Role of Governmental and Non-governmental Institutions in Local and Regional Economic Development in Western Europe*, Prague: Department of Social Geography and Regional Development, Charles University.

Sýkora, L. (1994b) 'Local urban restructuring as a mirror of globalization processes: Prague in 1990s', *Urban Studies* 31(7): 1149–1166.

Sýkora, L. (1995) 'Prague', in J. Berry and S. McGreal (eds) *European Cities, Planning Systems and Property Markets*, London: E & FN Spon, 321–344.

Sýkora, L. (1996) 'The Czech Republic', in P. Balchin (ed.) *Housing Policy in Europe*, London: Routledge, 272–288.

Sýkora, L. and I. Šimoníčková (1994) 'From totalitarian urban managerialism to a liberalized real estate market: Prague's transformations in the early 1990s', in M. Barlow, P. Dostál and M. Hampl (eds) *Development and Administration of Prague*, Amsterdam: Universiteit van Amsterdam: 47–72.

Szul, R. and Mync, A. (1997) 'The path towards the European integration: the case of Poland', *European Spatial Research and Policy* 4, 1: 5–36.

Teixidor, I.F. and Hebbert, M. (1982) 'Regional planning in Spain and the transition to democracy', in R. Hudson, and J. Lewis, (eds) *Regional Planning in Europe*, London.

Tewdwr-Jones, M. (ed.) (1996) *British Planning in Transition*, London: UCL Press.

Thatcher, M. (1988) *Britain and Europe*, London: Conservative Political Centre.

The European, 19–25 June 1997: 3 'Enlargement Opens a Can of Worms', Victor Smart (ed.).

Thomson, S. (1985) 'Local plans preparation: review of Scottish experience', paper presented to the research in land-use planning seminar, Oxford: Oxford Polytechnic.

Tomkins, J. and Twomey, J. (1994) 'Regional policy' in F. McDonald and S. Dearden, *European Economic Integration* (2nd edition) London: Longman.

Tóth, J. (1993) 'Transformation tendencies of the Hungarian system of settlements', in Z. Hajdú (ed.) *Hungary: Society, State, Economy and Regional Structure in Transition*, Pécs: Centre for Regional Studies, 155–183.

Town and Country Planning Association (1989) Bridging the north-south divide, London: TCPA.

Tsoukalis, L. (1997) *The New European Economy Revisited*, Oxford: Oxford University Press.

Tyler, P. and Moore, B.C. (1988) *'Geographical Variations in Industrial Costs'. Scottish Journal of Political Economy*, 35.

UMK (1997) *Plan rozwoju miasta Krakowa na lata 1997–2001*, Kraków: Urzad miasta Krakowa.

UNDP (1996) *Poland'96 – Habitat and Human Development*, Warsaw: UNDP.

Ustawa z 7. lipca 1994 r. o zagospodarowaniu przestrzennym.

Van den Berg, L., Drewett, R., Klassen, L.H., Rossi, A. and Vijverberg, C.H.T. (1982) *Urban Europe: a Study of Growth and Decline*, London: Pergamon

Vanhove, N. and Klassen, L.H. (1980) *Regional Policy: a European Approach*, Farnborough: Saxon House.

Vasconcelos, L. and Reis, C. (1994) 'The Portuguese planning process and local development plans', in P. Healey (ed.) *Trends in Development Plan-making in European Planning Systems* working paper no. 42, Newcastle: Department of Town and Country Planning, University of Newcastle-upon-tyne.

Verchére, I. (1994) 'Capital Problems for the Regions', *The European*, 1–7 July 1994.

Virtanen, P. (1994) 'A review of development plan-making in Finland', in P. Healey (ed.) *Trends in Development Plan-making in European Planning Systems*, working paper no. 42, Newcastle: Department of Town and Country Planning, University of Newcastle-upon-Tyne.

Vogel, D. (1995) 'The making of EC environmental policy', in M. Ugur (ed.) *Policy Issues in the European Union*, Dartford: Greenwich University Press.

Wannop, U. and Cherry, G. (1994) 'The development of regional planning in the United Kingdom', *Planning Perspectives* 9,1.

Wannop, U. (1995) *The Regional Imperative: Regional Planning and Governance in Britain, Europe and the United States*, London: Jessica Kingsley Publishers.

Wassenhoven, L. (1984) 'Greece' in M. Wynn (ed.) *Planning and Urban Growth in Southern Europe*, London, Mansell.

Wecławowicz, G. (1996) *Contemporary Poland: Space and Society*, Changing Eastern Europe Series 4, London: UCL Press.

Wegener, M. (1995) 'The changing urban hierarchy in Europe', in J. Brotchie, M. Batty, E. Blakely, P. Hall, and P. Newton, *Cities in Competition: Production and Sustainability for the 21st century*, Melbourne: Longman.

Welsh Economic Planning Council (1967) *Wales: the Way Ahead*, London: HMSO.

Welsh Office (1967) *Local Government in Wales*, Cardiff: HMSO.

Welsh Office (1990) *Strategic Planning Guidance, Structure Plans and the Content of Development Plans* (ppg15 Wales) Cardiff: Welsh Office.

Welsh Office (1992) *Development Plans and Strategic Planning Guidance in Wales* (ppg12 Wales) Cardiff: Welsh Office.

Welsh Office (1997) *A Voice for Wales*, Cardiff: The Stationery Office.

Wiehler, F. and Stumm, T. (1995) 'The power of regional and local authorities and their role in the European Union', *European Planning Studies*, 3, 2.

Williams, A.M. (1984) 'Portugal' in M. Wynn (ed.) *Planning and Urban Growth in Southern Europe*, London: Mansell.

Williams, A.M. (1987) *The Western European Economy: a geography of post-war development*, London: Hutchinson.

Williams, A.M. (1994) *The European Community*, Oxford: Blackwell.

Williams, R.H. (1996) *European Union Spatial Policy and Planning*, Paul Chapman Publishing, London.

Williams, A.M. (1997) *Western European Economy: a Geography of Post-war development*, London: Hutchinson.

Wynn, M. (ed.) (1984) *Planning and Urban Growth in Southern Europe*, London: Mansell.

Yuill, D., Bachtler, I. and Wishlade, F. (1996) *European Regional Incentives*, 16th edition, London: Bowker Saur.

Zoványi, G. (1986) 'Structural change in a system of urban places: the 20th-century evolution of Hungary's urban settlement network', *Regional Studies* 20, 1: 47–71.

Zoványi, G. (1989) 'The evolution of a national urban development strategy in Hungary', *Environment and Planning A* 21: 333–347.

Index